An introduction to
electromagnetic wave propagation and antennas

An introduction to electromagnetic wave propagation and antennas

Shane Cloude

First published in 1995 by UCL Press

UCL Press Limited
University College London
Gower Street
London WC1E 6BT

The name of University College London (UCL) is a registered trade mark
used by UCL Press with the consent of the owner.

British Library Cataloguing-in-Publication Data
A catalogue record for this book is available from the British Library.

ISBNs:
1-85728-240-X HB
1-85728-241-8 PB

Typeset in Sabon and Gill Sans.
Printed and bound by
Biddles Ltd., Guildford and King's Lynn, England.

Dedicated to the memory of John Bothwell

Contents

Preface

This book is intended as an introductory text on electromagnetic waves and antennas. It is aimed at second- or third-year university students in electrical and electronic engineering as well as in physics and applied mathematics. The structure and content follow a pattern familiar to many books of this kind; starting from Maxwell's equations in integral form we develop the wave equation and show how waves are generated by accelerating charge and propagate with the speed of light. This then leads us to a simplified quantitative consideration of wire and aperture antennas.

An unusual and hopefully useful feature of this book is the concentration on the use of time domain methods for a description of fundamental electromagnetic phenomena. This concentration on the time domain is deliberate and based on a firm belief that such methods offer a clarity and simplicity to introductory electromagnetic wave theory which is just not available using more conventional routes. That is not to say that the conventional approach based on rigorous vector calculus and formal solutions to the wave equation is not required at some stage. The main motive for this book was to provide as simple as possible an introduction to the difficult concepts and topics of electromagnetics, after which students will hopefully elect to fill in details by taking more advanced courses in mathematics and field theory. There are already many excellent books available for teaching electromagnetics at this advanced level (see the reading lists given at the end of each chapter) and this book does not seek to replace these, but merely to offer a treatment of the topics of antennas and wave propagation which will hopefully suit a wider range of student background and motivation.

With this in mind, this book requires only some background in introductory calculus and a familiarity with the basic techniques of integration and differentiation of multi-variable functions. While some formal proofs are supplied using the full vector calculus they are not central to the main development of the text and could easily be ignored on first use.

The book assumes that the students have some awareness of the basic equations of electromagnetism such as the laws of Ampère, Coulomb and Faraday and, using these as a basis, develops the topics in several important themes.

The first is the development of the wave equation, independently of its use in electromagnetics. This development follows experience with many students whose first introduction to the wave equation usually follows application of the machinery of vector calculus to Maxwell's equations. At this stage they very often cannot "see the wood for the trees" and are unable to extract the inherent simplicity of the wave equation.

This is particularly important given the modern trend towards using computational methods for modelling wave propagation, where techniques from a diverse range of physical disciplines are used. It is important therefore that students know which of these techniques can be used in electromagnetics and are also sensitive to situations where electromagnetic phenomena need special attention (such as in the case of wave polarization).

The second important theme is an honest attempt to answer a question often posed by students when they first meet the topic of electromagnetic radiation. How can open circuited wires radiate energy? The formal solution of fields around a Hertz dipole offers quantitative solace, but little in the way of physical understanding. In this book much attention is given to this theme, using as far as possible the mathematics of linear systems theory and impulse response, which engineering students will be familiar with from other courses.

The simple accelerating charge models are shown to be consistent with the Maxwell equations and are used as far as possible to develop simplified but useful models for a range of antenna types. The use of such accelerating charge models is not new, but as far as I am aware, they have not been pursued quite so far in the teaching of antenna concepts.

The third important theme is the use of numerical methods for solving electromagnetic wave problems. Methods of computational electromagnetics (CEM) are now widely used in research and in engineering product evaluation and development and have been the subject of several recently published advanced textbooks. In this text they are included in a supporting role, in the belief that when students face the problems of making numerical approximations to derivatives and integrals they see more clearly the physical basis for the equations.

For this reason the chapter on numerical methods is not intended as a comprehensive survey of all CEM techniques. The numerical methods considered are limited to the time domain. These not only support the time-domain analysis of other chapters but also provide an introduction to methods that are widely used at advanced levels of electromagnetic engineering such as the finite-difference and integral-equation methods. Example programs are provided in an Appendix B. These programs are written using the MATLAB programming language, but could be easily re-written in some other form. These provide a set of demonstrations and examples to be used by teachers or students themselves, depending on course structure.

The fourth important theme is the concept of aperture antennas. These are not developed using the traditional concepts of geometrical and physical optics, but are incorporated with the accelerating charge models by employing the concept of equivalent currents. This is usually a difficult idea for students to understand at first, but in the time domain it is clear how such an idea fits with the basic physical structure of the Maxwell equations. This will then give students some appreciation of the basic performance of horn and dish antennas for microwave communications and radar. No formal mention is made of system parameters related to the gain, effective area or directivity of antennas, although several related themes are developed in the exercises. Again it is believed that once students have grasped the physical principles, such secondary concepts can be easily introduced at a later date.

The final theme is that of electromagnetic scattering and radar. This advanced topic is not usually covered in introductory texts on electromagnetic waves, but is included here for two important reasons. Firstly, it employs the concepts of charge acceleration and equivalent currents to prove the simple but very important Kennaugh–Cosgriff formula for backscatter. The idea of generating an approximation to the impulse response of the back-scattered field is in harmony with earlier chapters. The second reason for incorporating an introduction to scattering is the increasing importance of this topic in electrical engineering. It has always been of interest in the physics of light and electron propagation but has recently become of great importance in the propagation of radio waves in the context of satellite communications and radar remote sensing.

While writing this book, with its bias towards time-domain techniques, I could not help some feelings of guilt at making no reference to Fourier- or Laplace-transform based techniques. After all, practising engineers make extensive use of such methods for quantitative calculations in electromagnetics. To alleviate these feelings somewhat, Appendix C shows how such methods relate to the time-domain formulations outlined in the text. However, it must be reiterated that while such transform methods are invaluable for *quantitative* electromagnetics, it is my firm belief that the time domain provides the best format for introducing students to the many and varied phenomena of electromagnetic waves and antennas.

With this in mind, I hope you, the reader, gain something of a new perspective from this book and, if after reading it, you feel a little easier with the concepts of electromagnetic waves, then all the hard work will have been worthwhile.

Shane Cloude

Acknowledgements

The subject matter of this book arose from a series of lectures given at the University of York, England, between 1991 and 1993. Thanks to all the students on those courses whose enthusiasm and searching questions motivated this project.

Thanks also to the York Electronics Centre, who supported the idea of industrial lectures on computational methods. Many of the ideas in this book were developed as a result of organizing courses for industrial engineers and scientists who attended the centre for intensive training courses on numerical modelling.

Special thanks go to Dr Alec Milne and Dr Paul Smith for sharing their insight into time-domain numerical methods.

Glossary of symbols

$\mathbf{E}$ electric field intensity vector in volts/metre (V/m)
$\mathbf{B}$ magnetic flux density vector in Tesla (T)
$\mathbf{D}$ electric flux density (displacement) vector in Coulombs/metre^2 (Cm^{-2})
$\mathbf{H}$ magnetic field intensity vector in Ampère/metre (Am^{-1})
ε permittivity in Farads/metre (Fm^{-1})
ε_0 permittivity of free space, $1/36\pi \times 10^{-9}\,Fm^{-1}$
ε_r relative permittivity, $\varepsilon/\varepsilon_0$
μ permeability in Henrys/metre (Hm^{-1})
μ_0 permeability of free space, $4\pi \times 10^{-7}\,Hm^{-1}$
μ_r relative permeability, μ/μ_0
c velocity of light in free space in metres/second (ms^{-1}), $2.998 \times 10^8\,ms^{-1}$
v wave velocity in metres/second (ms^{-1})
n refractive index, c/v
σ electrical conductivity in Siemens/metre (Sm^{-1})
q electric charge in Coulombs (C)
ρ electric volume charge density in Coulombs/metre^3 (Cm^{-3})
I current in Ampères (A)
$\mathbf{J}$ current density vector in Ampères/metre^2 (Am^{-2})
$\mathbf{J}_{ms}$ equivalent magnetic current density vector in Volts/metre^2 (Vm^{-2})
$\mathbf{J}_{es}$ equivalent electric current density vector in Ampères/metre^2 (Am^{-2})
R resistance in Ohms (Ω)
Z impedance (Ω)
Z_0 impedance of free space, $377\ \Omega$
L inductance in Henrys (H)
C capacitance in Farads (F)
λ wavelength in metres (m)
k wave-number in radians/metre ($rad\,m^{-1}$)
ω angular frequency in radians/second ($rad\,s^{-1}$)
f frequency in Hertz (Hz)
g pulse-width parameter for Gaussian pulse in seconds^{-1} (s^{-1})
t time in seconds (s)
$\mathbf{r}$ position vector, $x\mathbf{i} + y\mathbf{j} + z\mathbf{k}$
$\mathbf{i}, \mathbf{j}, \mathbf{k}$ Cartesian unit vectors
$h(t)$ impulse response function
$\delta(t)$ Dirac delta function
$G(\mathbf{r}, t)$ Green's function

$f*g$ convolution of functions f and g

∇ del, partial differentiation operator, $(\partial/\partial x, \partial/\partial y, \partial/\partial z)$

D1   singlet function, $\delta(t)$

D2   doublet function (first derivative of D1)

D3   triplet function (second derivative of D1)

Chapter 1

Wave motion
and the wave equation

1.1 Introduction to wave motion

Waves are ubiquitous in nature. We are familiar with many of their physical properties from our everyday experience. Waves on the surface of water, and acoustic or sound waves, are familiar physical phenomena.

In this book we look at the properties of a more exotic kind of wave motion; namely electromagnetic waves. Before studying their properties in any detail we first discuss waves in a general mathematical context. The main aim is to identify the important connections between different types of wave motion and to encompass them in a single mathematical equation, i.e. the wave equation, which we can then relate to the equations of electromagnetics.

It is important to realize that what follows applies to any kind of wave: from the exotic, such as electromagnetic waves, to the less interesting but more entertaining Mexican waves witnessed at sports arenas around the world. We aim to develop a single description of all types of wave, so that when we discuss some new physical process (such as the interaction of electric and magnetic fields) we will know how to check the governing equations to see what kinds of wave motion are possible.

Furthermore, it has recently become feasible to model complicated wave motions using computers and to visualize the dynamic wave interactions using sophisticated computer graphics. We shall discuss some of these methods in later chapters, but for now note that they too rely on the general mathematical descriptions that we are about to develop.

Visible light is an example of an electromagnetic wave, as are the radio and television signals we increasingly rely on in our information technology age. However, we do not normally perceive light as a wave motion in the same way as sound and water waves. There are two very good reasons for this; the first is that the velocity of light in our atmosphere is much higher than the velocity of other waves, such as sound (we shall show later that light travels with a velocity of $3 \times 10^8 \, \text{ms}^{-1}$ or 186,000 miles per second). Therefore we are not so aware of the delay caused by light to take a small but finite time to propagate from point A to point B in space as we are, for example, for sound waves (think only of seeing a flash of lightning in a thunderstorm long before you hear the thunder, despite the fact they have the same point of origin in space)(Fig. 1.1). However, we shall see that the existence of this very small time delay is critical in our understanding of how antennas can be designed to radiate energy efficiently.

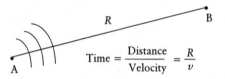

$$\text{Time} = \frac{\text{Distance}}{\text{Velocity}} = \frac{R}{v}$$

Figure 1.1 Basic wave propagation.

The second important reason we do not perceive light as a wave in "ordinary" experience is that light corresponds to waves with wavelengths which are very small compared to the scale of objects we encounter every day. Light has a wavelength of around $0.5\,\mu\text{m}$ ($1\,\mu\text{m} = 10^{-6}\,\text{m}$) and the interaction of light with everyday objects, which can be several million times larger than a wavelength, is very different from the interaction of sound waves, whose wavelengths in air are much longer (the speed of sound is $344\,\text{m s}^{-1}$ and so audible sounds have wavelengths of around $0.1\,\text{m}$).

There are five basic processes to be considered when studying wave motion:

- Generation
- Propagation
- Reflection
- Diffraction
- Scattering

These elementary processes are illustrated in Figure 1.2. In this book we examine the formulation and physical properties of these important wave phenomena in the context of electromagnetics.

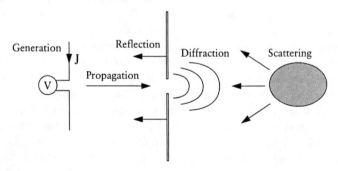

Figure 1.2 Summary of the five basic wave processes.

We shall see that all the apparently striking physical differences between light, water and sound waves are only a matter of scale and that light, like all waves, can be studied using a single mathematical equation.

There are two important physical ideas about waves we need to develop before we can generate a mathematical model:

- The first is that waves travel as large-scale coherent structures with wave speed v (Fig. 1.3). This is a familiar concept in our everyday experience of waves and will lead us to discuss the principle of causality, whereby if we effect some change at position A it will always take a finite time for this change to be communicated to a physically separate point B (the time will be given by the distance between the points A and B divided by the velocity v). This will be crucial in our development of radiation from accelerating electric charge. Note that, essentially, the concept of wave speed permits us to relate the spatial and time variations of a wave, i.e. the speed tells us how many distance units the wave moves in a given number of time units.

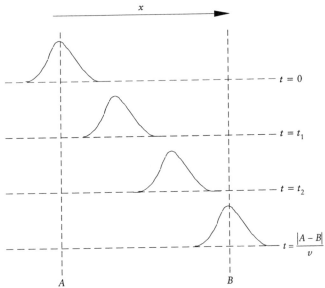

Figure 1.3 A wave as a moving coherent structure.

- A second key idea about waves is that they are simplest to describe on a "local" or small-scale level. This is an important idea and can best be illustrated by analogy with a Mexican wave. Although the Mexican wave seems to involve the coherent motion of a large number of people, it is clear to those who have been involved, that the "trigger" for an individual to react is entirely spatially and temporally local, i.e. the individual reacts at the appropriate time and space point simply by waiting until his or her neighbours react. There is no way you have to judge your reaction based on the timing and position of the original disturbance.

This simple idea of wave motion being communicated locally in space and time is a key one and leads us to describe wave motion mathematically in terms of differential equations. These equations tell us about differences between functions on a local level and can be solved to provide mathematical functions which satisfy these local rules.

Consider, for example, the propagation of sound waves away from a loudspeaker. The sequence of events may be represented schematically as shown in Figure 1.4.

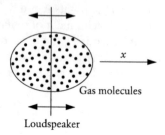

Figure 1.4 Basic mechanisms of sound-wave generation.

The waves are generated by an initial disturbance which is the physical movement of the loudspeaker cone in the x-direction. This causes the following chain of events:

- the gas (air) around the cone moves *locally* under the action of a force
- a local movement of gas molecules causes a corresponding local change in density (i.e. particles per unit volume)
- this change in density corresponds to a local change in pressure
- pressure inequalities generate gas motion (a physical law)

after which we see that we return to the second stage. In this way the wave propagates away from its initial source (the cone) as a feedback loop of physical processes, as summarized in Figure 1.5.

We shall use this feedback process as a paradigm for the identification of wave motion: we shall find in the next chapter a similar set of feedback mechanisms which can be used to explain the generation and propagation of electromagnetic waves.

Of course, having identified these mechanisms we can then ask questions about how the various stages depend on each other. For example, how fast do the gas particles move for a given pressure inequality? We shall not pursue such arguments for sound waves but will concentrate instead on the special case of electromagnetic waves

The identification of such a feedback process permits us to quantify important properties such as the velocity of wave propagation and the efficiency of

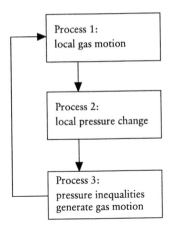

Figure 1.5 Feedback processes underlying sound-wave propagation.

the source element. In the example above we used a loudspeaker, which is carefully designed to maximize the coupling of energy from the movement of the cone into the launching of waves in space. Such an element in the electromagnetic wave context is called an "antenna", and the rules for designing efficient sources are the subject of the next chapter.

1.2   The wave equation

The time has now come for us to consider a mathematical formulation of these ideas. We begin by defining a function of position and time $f(r,t)$ where $r = x\mathbf{i} + y\mathbf{j} + z\mathbf{k}$, $\mathbf{i}$, $\mathbf{j}$ and $\mathbf{k}$ are unit vectors and x, y and z are the components of the position vector of a point in space. The function f can be used to represent any of a number of physical quantities such as the density of gas, height of a water wave or electric or magnetic field strength.

Our first key idea about waves implies that for our function to represent a wave, its spatial and temporal variation must be related by the velocity of wave propagation (otherwise it would not propagate as a coherent structure). In other words, the space and time dependence must be of the form $x \pm vt$ (where the choice of sign depends on the wave direction).

If we restrict attention to a wave propagating in one spatial dimension (the positive x direction) with velocity v, then it follows that our function must be of the form (see Problem 1.3)

$$f(x,t) = f(x - vt) \tag{1.1}$$

Our second principle leads us to try and obtain a local relation which our function f must satisfy. We can do this by generating the partial derivatives of the function f with respect to time and space, defined as

$$\frac{\partial f}{\partial x} = f_x \qquad \frac{\partial f}{\partial t} = f_t \tag{1.2}$$

where we have used a shorthand subscript notation for partial derivatives. It follows from the relation between space, time and velocity (Eq. (1.1)) that these two partial derivatives must satisfy the following differential equation

$$f_t + v f_x = 0 \tag{1.3}$$

which is a fundamental first-order wave equation that we shall call the *advection equation*. Functions which satisfy this local constraint represent waves moving in the positive x direction.

A similar argument provides us with a second form of this equation for waves travelling in the negative x direction. These functions must be of the general form $f(x + vt)$ and satisfy the following advection equation:

$$f_t - v f_x = 0 \tag{1.4}$$

A very important example of a wave function is generated from the sinusoidal function, defined as $f(x) = A \sin(kx)$, where k is a positive constant. This function is periodic with period given by $2\pi/k$ (Fig. 1.6).

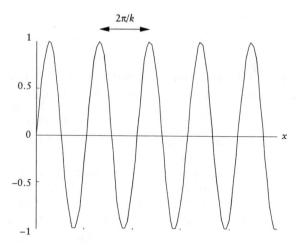

Figure 1.6 The sinusoidal function $\sin(kx)$.

Note that as it stands this function does not represent a wave. We can turn it into a wave function simply by replacing x by $(x - vt)$ to obtain

$$f(x,t) = A\sin\left[k(x - vt)\right] = A\sin(kx - \omega t) \tag{1.5}$$

from which we see that the wave velocity v is defined as

$$v = \frac{\omega}{k} \qquad \omega = 2\pi f \qquad k = \frac{2\pi}{\lambda} \qquad (1.6)$$

where ω is the angular frequency, f is the frequency (cycles per second, or Hertz) and λ is the wavelength of the sinusoid. We shall make extensive use of this special wave function when discussing the radiation and impedance properties of antennas.

While the first-order advection equations (Eqs (1.3) and (1.4)) can be used to model wave propagation in one spatial direction, we can combine them to generate a single second-order equation which models waves moving in either direction i.e. functions which are of the form $f(x \pm vt)$. This will then allow us to model the important phenomenon of wave reflection, whereby the wave direction is reversed.

Waves travelling in the $\pm x$ direction must satisfy the following product of advection equations as

$$(f_t + vf_x)(f_t - vf_x) = f_{tt} - v^2 f_{xx} = 0 \qquad (1.7)$$

This second-order differential equation is called the "wave equation" and is our generic mathematical description of wave motion. The key task of relating the physics of a process to the parameters in this equation involves a detailed understanding of the particular wave mechanism, but all waves satisfy an equation of this type.

As an example, we show in Appendix A how voltage and current waves can exist on electrical transmission lines. Electrical waves such as these are always guided by some conducting material (such as a pair of wires). We shall see later, however, that a more exotic form of electric and magnetic wave can exist in unbounded space.

We can easily extend these ideas to model waves propagating in any direction in space by generalizing the one-dimensional form of Eq. (1.7) to include spatial variations in the y and z directions to obtain the basic form of the three dimensional wave equation as

$$f_{tt} - v^2(f_{xx} + f_{yy} + f_{zz}) = g \qquad (1.8)$$

where we have also included a function $g(\mathbf{r}, t)$, which is a general source term, telling us how the waves are generated in the first place. In our acoustic wave example (Fig. 1.4), g would model the effect of the loudspeaker.

When $g \neq 0$ we have an *inhomogeneous* wave equation which tells us how the waves are generated and how they interact with objects in the processes of diffraction and scattering. When $g = 0$ we have a *homogeneous* equation which tells us how waves propagate in regions far away from any sources. Both types of equation are of use when studying a new wave motion.

The task of studying possible wave motions in a general system then reduces to trying to formulate the governing physical equations as a form of the wave equation relating second spatial and temporal partial derivatives. The coefficient of the time derivative then tells us about the velocity of such waves and the source term tells us how they originate. By understanding the physical origin of these terms we can then engineer structures which are efficient at generating such waves.

Boundary conditions

As well as satisfying the wave equation, wave solutions must also satisfy some boundary and initial conditions. These conditions are often imposed by the physics of the system (for example at time $t = 0$ we often assume that all currents in a system are zero on the physical premise that without a drive source there can be no currents flowing). The zero initial conditions will permit us to solve wave problems numerically by employing a marching-in-time procedure to the wave equation.

Alternatively, boundary conditions are often deliberately imposed in order to force the currents to be zero for all time at selected points in space. We shall see that these boundary points are very important in that they generate reflections of current waves which give rise to efficient generation of radiation.

Note that although the wave equation is a general description of all types of wave, it tells us nothing about the physics underlying a given wave system; to understand a particular system we must always turn to the governing physical equations. In the case of electromagnetics these are called the "Maxwell field equations" or simply "Maxwell's equations".

1.3 Maxwell's equations

There are two primary methods of formulating the equations of electromagnetics: the simplest and most commonly used approach is to use a large-scale or integral form for Gauss's, Ampère's and Faraday's laws, where we express the fields in terms of the sum (or integrals) over current and charge, the fundamental sources of electric and magnetic fields. In the second method we describe these same relationships in small-scale or differential equation form. Clearly, from our discussion above, this latter approach is preferred for an investigation of wave properties. Unfortunately, to achieve this small-scale version requires some results from vector calculus. In this section we develop this small-scale approach to the equations of electromagnetism.

The equations governing the time and spatial evolution of electric and magnetic fields were developed by a number of people during the nineteenth century, but a Scottish physicist, James Clerk Maxwell (1831–1879), is attributed with their first complete formulation in small-scale or differential form. For this reason, the equations are known collectively as the "Maxwell field equations".

There are several ways of defining the basic field quantities used for a description of electromagnetic phenomena, but one of the simplest is to base their definition on the observation that they exert a force on a material particle carrying electric charge q coulombs and moving with velocity $v\,\mathrm{ms}^{-1}$, given in vector form by the Lorentz formula:

$$\mathbf{F} = q(\mathbf{E} + \mathbf{v} \times \mathbf{B}) \qquad (1.9)$$

The first part of this equation expresses the influence of the electric field $\mathbf{E}$ on the charge and shows that it generates a force on the particle in the same direction as the $\mathbf{E}$ vector. The second term is a concise vector statement of the complicated interaction of the magnetic field $\mathbf{B}$ with moving particles with velocity $\mathbf{v}$. The cross product tells us that the force is perpendicular to both the $\mathbf{v}$ and $\mathbf{B}$ vectors.

The vector quantities $\mathbf{E}$ (the electric field vector in volts per metre) and $\mathbf{B}$ (the magnetic field vector or flux density in webers per square metre or tesla) are "abstracted" from these observations and supposed to have an existence of their own in space and time. We then write them as vector functions $\mathbf{E}(\mathbf{r},t)$ and $\mathbf{B}(\mathbf{r},t)$. We must remember, however, that the only way these quantities interact with the world around us is through their influence on matter, given by the Lorentz force equation.

We can now envisage our electric and magnetic fields as existing everywhere in space, at each point of which we associate six numbers, the three electric field and three magnetic field components represented as (E_x, E_y, E_z, B_x, B_y, B_z). Such a concept is called a "vector field" and the study of such fields is an important branch of applied mathematics.

We now seek "local" spatial and temporal relationships between these quantities as we move from one point to another in space. We will find that while neither $\mathbf{E}$ nor $\mathbf{B}$ by themselves satisfy the local constraints of a wave equation, when considered together they interact in a subtle way which gives rise to a wave propagating in space. This wave is generated by a coupling of the $\mathbf{E}$ and $\mathbf{B}$ fields, hence the term "electromagnetic waves", to emphasize the fact that both field components are involved in the propagation mechanism.

Gauss's law for electric and magnetic fields

The first set of constraint equations on the vectors $\mathbf{E}$ and $\mathbf{B}$ can be obtained from Gauss's law for electric and magnetic fields. Carl Friedrich Gauss (1777–1855) was a German mathematician and a pioneer in the application of mathematics to electricity and magnetism. Gauss's law relates a mathematical quantity, the total flux of electric and magnetic fields through a closed surface, to the presence of electric charge inside the surface. The law implies the existence of isolated electric charges with like charges repelling and unlike ones attracting. For magnetic fields, the law states that the total flux is always zero, a result which is consistent with the observation that isolated magnetic charges (so-called "magnetic monopoles") do not exist in nature.

While the laws apply to static charges, we shall see that the "local" constraints they impose on the field vectors have one very important consequence for dynamic problems when the charges move and radiation occurs: Gauss's law imposes the constraint that electromagnetic waves must be transverse waves, i.e. they exhibit the phenomenon of wave polarization.

In integral form, Gauss's law has the following concise mathematical form

$$\iint_S \mathbf{E.ds} = \frac{1}{\varepsilon} \iiint_V \rho \, dv$$

$$\iint_S \mathbf{B.ds} = 0 \tag{1.10}$$

where S is a general closed surface enclosing a volume V (Fig. 1.7), ε is the electrical permittivity of the medium (assumed homogeneous), ρ is the electric charge density inside V, and the vector surface element $\mathbf{ds} = \mathbf{n} \, da$ where da is the scalar area of a small segment of surface and $\mathbf{n}$ is the local surface normal.

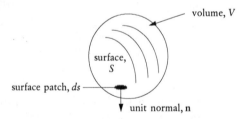

volume, V

surface, S

surface patch, ds

unit normal, n

Figure 1.7 General surface S enclosing a volume V in space.

These integral relations may be loosely interpreted as mathematical statements of the fact that field lines only ever terminate on free charges. Since there are no isolated magnetic poles in nature, the number of magnetic field lines entering a volume must equal the number leaving, which is why the surface integral for **B** always equals zero (the net flux is the difference between the number of field lines entering V and the number leaving). For electric fields this difference is equal to the total electric charge inside the volume V.

We can turn this large-scale form of Gauss's law into a small-scale law telling us more about the variation of **E** and **B** locally around a point in space by using a mathematical theorem called the *Divergence theorem*, which has the following mathematical form:

Divergence theorem

$$\iint_S \mathbf{A}.ds = \iiint_V \text{div } \mathbf{A} \, dv \qquad (1.11)$$

where $\mathbf{A}$ is some vector field and the surface S bounds the volume V. This integral relation states that the net flux through a surface is equivalent to the sum of the divergence of $\mathbf{A}$ through all the elementary volumes enclosed by S. This theorem will allow us to express Eq. (1.10) as a relation between volume integrals on both sides of the equation and so achieve a small-scale version of Gauss's law. However, before that we must ponder as to the physical significance of the divergence of a vector.

The divergence of a vector is a scalar (or single number), defined as a sum of partial derivatives:

$$\text{div } \mathbf{A} = \frac{\partial A_x}{\partial x} + \frac{\partial A_y}{\partial y} + \frac{\partial A_z}{\partial z} = \nabla.\mathbf{A} \qquad (1.12)$$

where we have introduced the "nabla" or "del" operator ∇, which is a vector with elements given by the partial derivative operators with respect to space co-ordinates, i.e.

$$\nabla = \left(\frac{\partial}{\partial x} \quad \frac{\partial}{\partial y} \quad \frac{\partial}{\partial z} \right) \qquad (1.13)$$

The divergence is then formed as the scalar product between ∇ and the vector $\mathbf{A}$. (Although we may consider ∇ as a vector and so, strictly, it should be represented in a bold typeface, for notational convenience we join with common practice and use ∇.)

We can establish the physical basis of the combination of partial derivatives represented by the divergence by considering the *flux* of a vector quantity $\mathbf{B}$ (the magnetic flux density for example) through an infinitesimal cube of dimensions dx, dy and dz as shown in Figure 1.8, where we show only one dimension of the cube of side dx. By definition of flux as flux density times

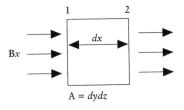

Figure 1.8 Flux through a small cube.

area, and using the concept of partial derivatives for small changes, we can write the flux through surfaces 1 and 2 as

$$\text{Flux from } 1 = -B_x \, dy \, dz$$

$$\text{Flux from } 2 = \left(B_x + \frac{\partial B_x}{\partial x} dx \right) dy \, dz$$

Note the negative sign in the flux from 1 and the definition of area of the cube face as $A = dy\,dz$. The net flux through the cube in the x direction is then the sum of these terms:

$$\text{Flux} = \frac{\partial B_x}{\partial x} dx \, dy \, dz = \frac{\partial B_x}{\partial x} dv$$

We can perform similar calculations in the y and z directions to obtain the total flux through the surface of the cube as

$$\iint \mathbf{B}.\mathbf{ds} = \left(\frac{\partial B_x}{\partial x} + \frac{\partial B_y}{\partial y} + \frac{\partial B_z}{\partial z} \right) dv = \nabla.\mathbf{B} \, dv \qquad (1.14)$$

The divergence theorem (Eq. (1.11)) generalizes this result for a volume of space constructed from an assembly of such elementary volumes. By using the divergence theorem we then obtain the following relations from Gauss's Law:

$$\iiint_V \text{div}\,\mathbf{E} \, dv = \frac{1}{\varepsilon} \iiint_V \rho \, dv$$

$$\iiint_V \text{div}\,\mathbf{B} \, dv = 0 \qquad (1.15)$$

which at last gives us some differential or small-scale constraints on the vectors $\mathbf{E}$ and $\mathbf{B}$. Since the volume V is arbitrary, it follows that the above relations apply at any point and hence we arrive at the first two of Maxwell's (small-scale) equations

Maxwell's equations from Gauss's law

$$\text{div}\,\mathbf{E} = \frac{\rho}{\varepsilon} \qquad\qquad \text{div}\,\mathbf{B} = 0 \qquad (1.16)$$

Thus, any theory we consider for the $\mathbf{E}$ and $\mathbf{B}$ vectors must be constrained such that these sums of partial derivatives apply.

Faraday's law of electromagnetic induction

In contrast to Gauss, Michael Faraday (1791–1867) was an English experimental physicist renowned not for his mathematical ability but for his experimental expertise in studying the dynamics of electric and magnetic fields. He is credited with the discovery of the principle of the electric motor and of the phenomenon of electromagnetic induction, the latter being of great importance in our discussion of wave motion.

Faraday's law states that a changing magnetic field generates an electric field. Although often applied to the generation of current flow in closed electrical circuits, it is important to realize that Faraday's law is a general result, applicable to any closed contour in space, whether or not a wire circuit is present.

In integral form it states that the line integral of the electric field E around a closed contour C (called the "circulation" of E) is equal to the time rate of change of the total magnetic flux through the contour. In mathematical form we have

$$\oint_C \mathbf{E}.\mathbf{dl} = -\frac{\partial}{\partial t} \iint_S \mathbf{B}.\mathbf{ds} \tag{1.17}$$

where the circle on the integral sign is there to remind us that the integration is around a closed contour and S is a surface spanning C as shown in Figure 1.9

Figure 1.9   Contour C around a closed surface S.

This again is a large-scale law which we must convert into small-scale or differential form. We can do this by trying to express the left-hand side of Eq. (1.17) as a surface integral. To do this we use Stokes' theorem, which permits us to express a line integral around a closed contour as a surface integral of the normal component of a new small-scale quantity, the curl of a vector field, as

Stokes' theorem

$$\oint_C \mathbf{A.dl} = \iint_S \mathrm{curl}\,\mathbf{A.ds} \qquad (1.18)$$

where the curl of a vector $\mathbf{A}$ is a combination of first-order spatial partial derivatives defined in Cartesian co-ordinates by the following determinant

$$\mathrm{curl}\,\mathbf{A} = \nabla \times \mathbf{A} = \begin{vmatrix} \mathbf{i} & \mathbf{j} & \mathbf{k} \\ \dfrac{\partial}{\partial x} & \dfrac{\partial}{\partial y} & \dfrac{\partial}{\partial z} \\ A_x & A_y & A_z \end{vmatrix} \qquad (1.19)$$

Stokes' theorem has a particularly simple geometrical interpretation if we imagine a planar surface constructed from small elementary surface elements as shown in Figure 1.10.

The term curl $\mathbf{A}$ is defined as the circulation of a vector $\mathbf{A}$ around a small surface element in the limit as the enclosed area tends to zero. Hence we obtain the result that the net circulation only depends on the integral around the boundary of the surface, since all interior points cancel, as shown in Figure 1.10.

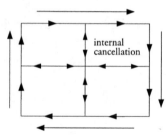

Figure 1.10 Planar surface showing internal cancellation in Stokes' theorem.

We can use this idea to demonstrate the importance of the particular combination of partial derivatives employed in defining the curl operation. Consider that the contour C is a small rectangular loop in the xy plane, as shown in Figure 1.11. The circulation of a vector field $\mathbf{E}$ around such a loop is then,

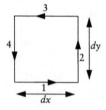

Figure 1.11 Elementary loop in space.

by definition, given by

$$\oint \mathbf{E} \cdot \mathbf{dl} = E_x(1)dx + E_y(2)dy - E_x(3)dx - E_y(4)dy \qquad (1.20)$$

where we have used the notation $E(1)$ to mean the value of $\mathbf{E}$ along edge 1, etc. Using partial derivatives and the fact that the loop is very small permits the following approximation

$$E_x(3) = E_x(1) + \frac{\partial E_x}{\partial y}dy \Rightarrow \left[E_x(1) - E_x(3)\right]dx = -\frac{\partial E_x}{\partial y}dx\,dy$$

Similarly, we have

$$\left[E_y(2) - E_y(4)\right]dy = \frac{\partial E_y}{\partial x}dx\,dy$$

The total circulation around the loop in the xy plane is then

$$\left(\frac{\partial E_y}{\partial x} - \frac{\partial E_x}{\partial y}\right)dx\,dy$$

The quantity in parentheses is identified as the z component of the vector defined by the cross product $\nabla \times \mathbf{E}$.

Similar expressions can be obtained for rectangular loops in the xz and yz planes to obtain the three vector components of the curl operator.

We can see then that this rather strange combination of partial derivatives called the "curl" arises naturally when describing the circulation of a vector field around small loops in space. It is fortunate that we can summarize these relationships using the short-hand notation of a vector product which tells us not only about the magnitude of the circulation but also its orientation in space.

If we apply Stokes' theorem to Faraday's law we obtain an equality between surface integrals which again (since the surface is arbitrary) we can generalize to points in space. There results the following small-scale form of Faraday's law.

Maxwell's equation from Faraday's law (Eq. (1.17))

$$\nabla \times \mathbf{E} = -\frac{\partial \mathbf{B}}{\partial t} \qquad (1.21)$$

Note that due to the cross product in the definition of the curl operation, the E field generated by a changing magnetic field is perpendicular to the source field $\mathbf{B}$.

Ampère's law

André-Marie Ampère (1775–1836) was a French physicist who was the first to name and study the subject of electromagnetism. Ampère's law relates the magnetic fields generated by currents and was developed to describe the force observed between current-carrying conductors. From the very earliest experimental investigations it was realized that such a relationship is much more complicated than that for electric fields. Ampère showed that these observations could be cast in concise mathematical form as an integral relation of the form

$$\oint_C \mathbf{B}.\mathbf{dl} = \mu \iint_S \mathbf{J}.\mathbf{ds} \tag{1.22}$$

where μ is the magnetic permeability, the surface S and the contour C are related as for Faraday's law and $\mathbf{J}$ is the current density (Ampères per square metre). Again by using Stokes' theorem to express the left-hand side as a surface integral we obtain the following small-scale differential relationship between the spatial derivatives of $\mathbf{B}$ and the current $\mathbf{J}$

Maxwell's equation from Ampère's law (Eq. (1.22) and Eq. (1.18))

$$\nabla \times \mathbf{B} = \mu \mathbf{J} \tag{1.23}$$

We now have four small-scale or differential equations relating the vectors $\mathbf{E}$ and $\mathbf{B}$ in space and time.

So far our derivations have been a small-scale reformulation of the laws of Ampère, Gauss and Faraday. Maxwell's contribution was that he realized that this set of equations are, as they stand, contrary to the law of conservation of charge. To address this problem he postulated the existence of an additional term in Ampère's law, the displacement current. We shall see that it is this extra term which is vital in providing the coupling necessary for the propagation of waves.

1.4 Displacement current and the conservation of charge

Maxwell realized that the small-scale version of the equations of electromagnetics, as outlined above, contradicted a fundamental physical law, i.e. the law of conservation of charge. This law states that charge can only be removed from a volume if it flows as a current through the surface enclosing that volume, i.e. there is no spontaneous creation or annihilation of charge.

In mathematical terms it can be formulated as a small-scale or differential relationship between the rate of change of charge in a volume and the divergence of the current density through a surface S enclosing V as

$$\nabla.\mathbf{J} = -\frac{\partial \rho}{\partial t} \tag{1.24}$$

where the left-hand side is the divergence of the current vector and is the total current flowing through S and the right-hand side is the rate of decrease in the charge inside V. For the conservation of charge these two must be equal.

The problem is that this differential equation contradicts the differential form of Ampère's law, which by itself implies that the divergence of current is always zero since (from Eq. (1.23))

$$\nabla.\mathbf{J} = \nabla.(\nabla \times \mathbf{B}) = 0 \tag{1.25}$$

where the last step follows from the basic vector identity (div curl $\mathbf{A} = 0$) for any vector field $\mathbf{A}$ (which can be proved by straightforward expansion using the definition of ∇ and the rules of partial differentiation: see Problem 1.7).

To resolve this issue, Maxwell started from the assumption that the law of conservation of charge was correct and tried to investigate the implications for the curl relation obtained from Ampère's law. He began by defining a new vector, the electric displacement $\mathbf{D}$ (see Eq. (1.26)).

Using Gauss's law we can then rewrite the equation of continuity as

$$\nabla.\mathbf{J} + \frac{\partial}{\partial t}(\nabla.\mathbf{D}) = 0$$

$$\mathbf{D} = \varepsilon\mathbf{E} \qquad \rho = \nabla.\mathbf{D} \tag{1.26}$$

$$\nabla.\left(\mathbf{J} + \frac{\partial \mathbf{D}}{\partial t}\right) = 0$$

and this becomes consistent with Ampère's law if we make the following addition to the right-hand side of Eq. (1.23)

$$\nabla \times \mathbf{B} = \mu\mathbf{J} + \varepsilon\mu\frac{\partial \mathbf{E}}{\partial t} \tag{1.27}$$

The new source term on the right-hand side of this equation generates a magnetic field $\mathbf{B}$ and hence can be considered as a kind of current. It is called the "displacement current". Note that the displacement current is no more than the time derivative of the electric field multiplied by a pair of material constants. This term will provide us with the coupling between $\mathbf{E}$ and $\mathbf{B}$ necessary to support wave propagation.

It is interesting to note that Maxwell discovered the displacement current term using theoretical arguments alone, whereas most of the other field equations had their origins in the experimental observations of Faraday, Ampère and others. The reason for this stems from the fact that the time derivative of electric field is multiplied by two fundamental material constants, the permittivity ε and permeability μ of the material. Importantly, these parameters are never zero, not even for vacuum where there are no material particles! Thus the displacement current is never zero, unless the electric field is unchanging with time.

Values for ε_0 and μ_0, the parameters of free space, are known from static field measurements (i.e. fields where there is no time variation) as follows:

$$\varepsilon_0 = 8.854 \times 10^{-12} \,\mathrm{F\,m^{-1}}$$

$$\mu_o = 4\pi \times 10^{-7} \,\mathrm{H\,m^{-1}}$$

The product of these two is a very small number (and, as we shall see, is related to the velocity of light). Clearly the contribution of the displacement term in vacuum is going to be very small (much too small to see experimentally) unless the time derivative of the electric field is large enough. When experiments were performed by Heinrich Hertz (1857–1894) to investigate the behaviour of sinusoidal fields at high frequencies, where the time derivative is large, the theoretical predictions of Maxwell were indeed confirmed. In particular, it was confirmed that the velocity of propagation of these Hertzian waves in vacuum, commonly denoted by c, was given by

$$c = \frac{1}{\sqrt{\varepsilon_0 \mu_0}} \tag{1.28}$$

which yields a value of $2.998 \times 10^8 \,\mathrm{m\,s^{-1}}$, which is the same as the known velocity of light. In this way light was first identified as an electromagnetic wave. Furthermore, if light is a wave composed of electric and magnetic fields then we should be able to use Maxwell's equations to generate a wave equation which will tell us more about how to generate the waves. We shall pursue this idea in Chapter 2.

1.5 The electromagnetic spectrum

Now we have a value for the speed of wave propagation in free space ("free" in the sense that it is devoid of material particles), we can relate the wavelength (the distance between peaks of the sinusoid) of sinusoidal waves to their frequency (the number of cycles per second). The basic engineering impact of this is that we can by various electrical means excite sinusoidal waves by forcing voltages and currents to oscillate with a fixed number of cycles per second and, knowing the velocity, we can then calculate the corresponding

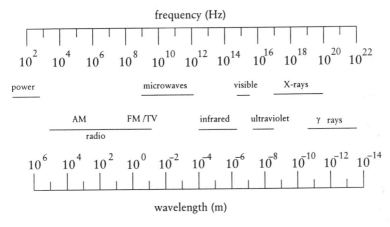

Figure 1.12 The electromagnetic wave spectrum.

wavelength in space. We shall see that the wavelength of the wave is important in determining how the wave propagates and interacts with structures to undergo diffraction and scattering.

The variation of f with λ is termed the electromagnetic spectrum and some important regions of this spectrum are indicated in Figure 1.12.

Concluding remarks: the complete set of Maxwell's equations

Note that we have used the short-hand "del" notation (∇) in our differential form of Maxwell's equations. We can also use this notation to write the three-dimensional wave equation in short-hand form, since

$$\frac{\partial^2}{\partial x^2} + \frac{\partial^2}{\partial y^2} + \frac{\partial^2}{\partial z^2} = \nabla.\nabla = \nabla^2 \tag{1.29}$$

where we have formed the scalar product of the del operator with itself. This operator arises a great deal in electromagnetics and is called the "Laplacian" (after the French astronomer and mathematician Pierre Simon Laplace (1749–1827)). Strictly it applies as an operator on scalar functions, but if we define $\nabla^2 \mathbf{A}$ as the vector generated by applying the Laplacian to each element of $\mathbf{A}$ then we obtain the wave equation in a short-hand form as

$$\nabla^2 \mathbf{A} - \frac{1}{v^2}\frac{\partial^2 \mathbf{A}}{\partial t^2} = \mathbf{g}(\mathbf{r},t) \tag{1.30}$$

where $\mathbf{A}$ is a vector field and $\mathbf{g}$ is a vector source term.

Table 1.1   The Maxwell field equations.

$\nabla . \mathbf{B} = 0$	Gauss's law for magnetic fields
$\nabla . \mathbf{E} = \dfrac{\rho}{\varepsilon}$	Gauss's law for electric fields
$\nabla \times \mathbf{B} = \mu \mathbf{J} + \varepsilon \mu \dfrac{\partial \mathbf{E}}{\partial t}$	Ampère's law + displacement current
$\nabla \times \mathbf{E} = -\dfrac{\partial \mathbf{B}}{\partial t}$	Faraday's law
$\nabla . \mathbf{J} + \dfrac{\partial \rho}{\partial t} = 0$	Continuity equation

We have now developed all the equations we need to study electromagnetic waves. We started out wanting to obtain small-scale or differential relationships between the field vectors **E** and **B** arising in electromagnetics and we now have them. The full set of Maxwell equations in differential form are shown in Table 1.1. These are all the equations we need to investigate electromagnetic wave motion. They tell us everything we need to know about how local interactions occur between the electric and magnetic fields. We shall now use these to investigate the physical properties of electromagnetic waves and show how the wave equation arises from Maxwell's equations.

Suggestions for further reading

Details of many of the ideas presented in this chapter can be found in standard texts on electromagnetics. Of the many available, the following are recommended for their clarity and variety of approach.

Born, M. & E. Wolf 1989. *Principles of optics*. Oxford: Pergamon Press.

Feynman, R., R. Leighton, M. Sands 1975. *Lectures on physics: volume 2*. New York: Addison Wesley.

Jones, D. S. 1989. *Acoustic and electromagnetic waves*. Oxford: Oxford Science Publications.

Kong, J. A. 1986. *Electromagnetic wave theory*. New York: Wiley.

Kraus, J. D. 1984. *Electromagnetics*, 3rd edn. New York: McGraw Hill.

Lee, K. 1984. *Principles of antenna theory*. New York: Wiley.

Longair, M. S. 1984. *Theoretical concepts in physics*, chaps 3 & 4. Cambridge: Cambridge University Press.

Marshall, S. V. & G. G. Skitek 1990, *Electromagnetic concepts and applications*, Englewood Cliffs, NJ: Prentice Hall.

Problems

1.1 Calculate the time it takes light to travel a) 1 m b) 1 km c) 1000 km d) the distance between the Earth and the Sun (149 600 000 km).

1.2 Calculate the length of wire required to make an antenna of half a wavelength of sinusoidal electromagnetic waves with frequencies of
    i) 20 kHz   (1 kilohertz $= 10^3$ cycles s^{-1})
    ii) 20 MHz   (1 megahertz $= 10^6$ cycles s^{-1})
    iii) 1 GHz   (1 gigahertz $= 10^9$ cycles s^{-1})
    iv) 1 THz   (1 terahertz $= 10^{12}$ cycles s^{-1}).

1.3 Use the chain rule to prove that $f(x,t) = \psi(x \pm vt)$ is a solution of the one dimensional wave equation.

1.4 Given the profile $\varphi(x,0) = 10 \exp(-x^2)$, write an expression for the corresponding progressive wave travelling in the positive x direction with a velocity of 10 ms^{-1}. Sketch the profiles at $t = 0$ s and $t = 0.5$ s.

1.5 Show that the expression $\varphi(x,t) = 10\exp[-(4x^2 + 6xt + 9t^2)]$ is a wave and calculate its velocity and direction of propagation. Verify that it is indeed a solution of the one-dimensional wave equation.

1.6 Which of the following functions satisfy the three-dimensional wave equation and hence represent waves propagating in space? For each wave identify the direction of propagation and the wave velocity:
    i) $e(r,t) = \cos(\omega t - kz)\,\mathbf{i}$
    ii) $e(r,t) = 3\cos(\omega t + kz)\,\mathbf{i} + \cos(\omega t + kz)\,\mathbf{j}$
    iii) $e(r,t) = \cos(\omega t - kx)\,\mathbf{j} + \sin(\omega t - kx)\,\mathbf{k}$
    iv) $e(r,t) = 6\cos(\omega t + kz)\,\mathbf{j} + 2\sin(\omega t + kz)\,\mathbf{k}$
If the function in i) represents an electric field, use the small-scale version of Maxwell's equations to calculate the corresponding magnetic field. Obtain an expression for the ratio of electric and magnetic field strengths in terms of c (the velocity of light).

1.7 Another important operation with the ∇ symbol is called the "gradient". The gradient of a scalar ϕ is defined as a vector with components given by directional partial derivatives as

$$\nabla\phi = \mathrm{grad}\,\phi = \frac{\partial\phi}{\partial x}\mathbf{i} + \frac{\partial\phi}{\partial y}\mathbf{j} + \frac{\partial\phi}{\partial z}\mathbf{k}$$

By using the rules of partial differentiation and the definition of the gradient, show that the two following vector identities are true for any vector field $\mathbf{A}$:

$$\nabla.(\nabla \times \mathbf{A}) = 0$$

$$\nabla \times (\nabla \times \mathbf{A}) = \nabla(\nabla.\mathbf{A}) - \nabla^2\mathbf{A}$$

$\mathrm{div}\,A = \nabla \cdot A$
$\mathrm{curl}\,A = \nabla \times A$
$\mathrm{grad}\,A = \nabla A$

21

Chapter 2
The generation of electromagnetic waves

One of the most difficult concepts to understand in electromagnetics is the physical basis for radiation from wire and aperture antennas. In Chapter 1 we considered the action of a loudspeaker as the source of acoustic waves. Here the mechanical movement of the source is a clear mechanism for generation. In the electromagnetic case, however, there is no obvious mechanical action and yet energy can still be made to radiate away from a source. Antenna engineering is concerned with the design and characterization of metal and dielectric structures which form *efficient* radiating elements. In this chapter we aim to outline the physical principles involved when designing such antennas.

We begin by describing the physical radiation and impedance properties of wire antennas such as dipoles, loops and helices, using a time-domain description of currents and fields around the wires. This time-domain approach yields clear physical insight into the properties of these antennas.

We first justify our use of time-domain models by deriving the convolution relationship between antenna excitation and radiated electric and magnetic fields. We then use Maxwell's equations to prove that radiated fields originate from accelerating charge and use this simple physical result to study the impedance and radiation properties of wire antennas.

2.1 The impulse response of linear systems

Although a rigorous quantitative analysis of antennas requires formal solutions to the Maxwell differential equations with appropriate boundary conditions, we note that, despite their complexity, Maxwell's equations form a linear system. By "linear" we mean that we can discuss their properties in terms of input/output relationships that have the following two important properties (the inputs are generally excitations due to current or voltage sources and the outputs will be the resultant electric and magnetic fields set up in space and time):

Property 1:
If we have an input function $x(\mathbf{r},t)$ which generates an output $y(\mathbf{r},t)$, then multiplying $x(\mathbf{r},t)$ by a constant α, generates a similarly scaled output $\alpha y(\mathbf{r},t)$.

Property 2:
If we have any two inputs $a(\mathbf{r},t)$ and $b(\mathbf{r},t)$ producing outputs $c(\mathbf{r},t)$ and $d(\mathbf{r},t)$ then when the input is a weighted sum of the two inputs, i.e. $S(\mathbf{r},t) = \alpha a(\mathbf{r},t) + \beta b(\mathbf{r},t)$, the output will have a similar form i.e. the output will be $\alpha c(\mathbf{r},t) + \beta d(\mathbf{r},t)$.

Property 2 is called the *superposition principle* and gives us the flexibility to choose special input functions and then form a whole new set of inputs via summation. This is a powerful concept in linear-systems theory and we shall make use of it in our discussion of Maxwell's equations. It will lead us to describe an antenna system in terms of a convolution integral, where the only unknown will be the system impulse response. We shall see that from Maxwell's equations we can obtain simple but effective approximations of the impulse response of a wide range of antenna systems. These approximations will permit us to describe many of the qualitative features of antennas and the engineering trade-offs involved in their design.

We first describe an arbitrary antenna as a linear system, with the input and output related as shown in Figure 2.1, where $x(t)$ is the input excitation to the antenna (in the form of voltage and current sources) and $y(t)$ is the observed radiated field. The attraction of this approach is that we do not need to consider the details of the internal structure of the system inside the box to derive a general relationship between the input and output.

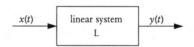

Figure 2.1 Linear system with single input and output.

By considering the antenna in this way as a "black box" with inputs and outputs, and using the superposition principle, we can relate the output to arbitrary input via the following convolution integral:

$$y(t) = \int_{-\infty}^{\infty} h(t-\tau)\,x(t)\,d\tau \qquad (2.1)$$

where the function $h(t)$ is called the "impulse response" of the linear system and is the response of the system to a special input, a Dirac delta function defined by $\delta(t) = 0$ except at $t = 0$ such that

$$\int_{-\infty}^{\infty} \delta(t)\,dt = 1 \quad \text{and} \quad \int f(t)\delta(t - t_b)\,dt = f(t_b) \qquad (2.2)$$

The Dirac function is specially constructed to have just these integral properties since, as we shall see, they are very useful in the analysis of linear systems.

We can prove the above convolution relation by considering the five stages of development shown in Figure 2.2. In this diagram, L is a linear system. In (a) we show how the impulse response $h(t)$ is defined as the system output with a delta function input. In (b) we consider the general case where $x(t)$ and $y(t)$ are arbitrary functions. In (c) and (d) we use the shift property of the delta function to show how the output at time t' is related to the product of input function and impulse response. Finally, in stage (e) we use the superposition principle to express the input as a sum of weighted delta functions and obtain our convolution integral as output.

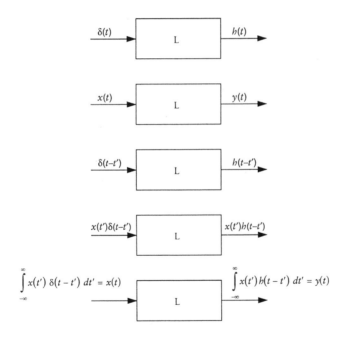

Figure 2.2 Stages involved in the development of the convolution integral.

In numerical and analytical time-domain problems we cannot represent the delta function exactly, since it effectively has zero duration and infinite amplitude. Nonetheless we can use a series of well-behaved functions to approximate this special kind of input. In most time-domain problems, we make use of the Gaussian function, which is defined as

$$v(t) = \frac{g}{\sqrt{\pi}} \exp\left[-g^2(t - t_b)^2\right] \Rightarrow \int_{-\infty}^{\infty} v(t)dt = 1 \qquad (2.3)$$

which in the limit as g tends to infinity, approximates the delta function. In numerical or experimental work it is sufficient to choose a value of g large enough to yield a pulse width much smaller than the time taken for light to traverse the antenna structure. The resulting response waveform then gives a good approximation to the system impulse response.

Note that the parameter t_b is a bias time to ensure that the pulse is causal, i.e. that its peak value occurs at time $t = t_b > 0$. The parameter g is related to the full width at half height (FWHH) of the Gaussian function (which is a measure of the pulse width in seconds) by

$$\exp\left(-g^2 t^2\right) = \frac{1}{2} \quad \Rightarrow \quad t = \frac{\sqrt{\ln 2}}{g}$$

$$2t = \text{FWHH} = \frac{2\sqrt{\ln 2}}{g} \approx \frac{1.665}{g} \qquad (2.4)$$

Figure 2.3 shows a Gaussian pulse with $g = 1.665 \times 10^9$, which corresponds to a FWHH of 1 ns. The bias time t_b is 5 ns. Note that the peak amplitude is very large; this is to ensure that the area under the pulse is unity.

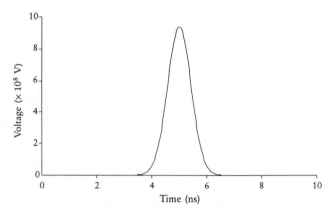

Figure 2.3 The Gaussian function as an approximation to the delta function.

In the remainder of this chapter we consider delta function excitation of wire antennas, with the understanding that for quantitative estimates, functions such as the Gaussian pulse must be used. We shall represent the delta function schematically as shown in Figure 2.4

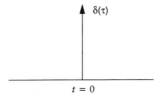

Figure 2.4 Schematic representation of the delta function.

The problem now remains to obtain estimates of $h(t)$, the antenna impulse response, for which we need to understand the physical behaviour of the linear system. We shall see that by using a few simple physical rules obtained from Maxwell's equations, we can obtain good approximations of $h(t)$ for a range of antenna types. This will permit us to describe the qualitative behaviour of such antennas, with the understanding that fully quantitative estimates require numerical or analytical solutions to Maxwell's equations.

Before proceeding we note that if the function $h(t)$ can itself be expressed as a sum of delayed delta functions of the form

$$h(t) = \sum_{k=0}^{\infty} a_k \, \delta(t - t_k) \tag{2.5}$$

then the convolution integral to determine the response to arbitrary excitation is easily evaluated as

$$y(t) = \int_{-\infty}^{\infty} \sum a_k \delta(t - t_k - \tau) x(\tau) d\tau = \sum a_k x(t_k) \tag{2.6}$$

In particular if $x(t) = \sin(\omega t)$ then we obtain the sinusoidal response of the antenna.

We can extend this idea to formally define the convolution of two discrete sequences f and g written as f*g and defined as the sum

$$c(x) = \sum_{m=0}^{M-1} f(m) g(x - m) = f(x) * g(x)$$

$$f(x) = (f_0 \; f_1 \; f_2 \cdots f_{A-1}) \tag{2.7}$$

$$g(x) = (g_0 \; g_1 \; g_2 \cdots g_{B-1})$$

For good estimates of the convolution we must choose $M > A + B - 1$. The following simple example will illustrate this.

Example: The result of the convolution of the sequence (2 1 3 4) with (1 4 3 1) is the sequence (2, 9, 13, 21, 26, 15, 4) (Fig. 2.5).

$$
\begin{array}{ccccccccl}
 & & 2 & 1 & 3 & 4 & & & \Rightarrow \\
1 & 3 & 4 & 1 & & & & & \Rightarrow & 2 \\
 & 1 & 3 & 4 & 1 & & & & \Rightarrow & 9 \\
 & & 1 & 3 & 4 & 1 & & & \Rightarrow & 13 \\
 & & & 1 & 3 & 4 & 1 & & \Rightarrow & 21 \\
 & & & & 1 & 3 & 4 & 1 & \Rightarrow & 26 \\
 & & & & & 1 & 3 & 4 & 1 & \Rightarrow & 15 \\
 & & & & & & 1 & 3 & 4 & 1 & \Rightarrow & 4
\end{array}
$$

Figure 2.5 Numerical convolution of two sequences.

We now consider the physical rules required to obtain estimates of the impulse response $h(t)$ for antenna problems.

2.2 Radiation and Maxwell's equations

The governing equations for the electric and magnetic fields generated by currents with density $J\,A\,m^{-2}$ and charge with density $\rho\,C\,m^{-3}$, were derived in Chapter 1 as

$$
\nabla \times E = -\frac{\partial B}{\partial t}
$$

$$
\nabla \times B = \mu J + \varepsilon\mu\frac{\partial E}{\partial t}
$$

$$
\nabla.B = 0
$$

$$
\nabla.E = \frac{\rho}{\varepsilon}
$$

where all quantities are functions of space and time, e.g. $E(r,t)$. (For the sake of clarity, we shall omit this explicit dependence from the equations). We now consider the following proposition about electromagnetic waves:

Proposition 1: It follows from Maxwell's equations that electromagnetic radiation occurs whenever electric charge accelerates.

We first prove this proposition formally from Maxwell's equations using some standard vector identities and then illustrate how it arises more physically by considering the problem of an accelerating isolated charge.

Formal proof of Proposition 1:

To prove that electromagnetic waves exist we need to formulate Maxwell's equations as a wave equation of the form

$$\nabla^2 \psi - \frac{1}{v^2} \frac{\partial^2 \psi}{\partial t^2} = g \tag{2.8}$$

where ψ is one of our field quantities and the source term $g(r,t)$ tells us how such waves arise. According to our proposition this will involve the second time derivative of the electric charge. The coefficient of the partial second time derivative of ψ will then tell us about the velocity of such waves.

We can obtain a wave equation from Maxwell's equations in four stages.

Stage 1:

Take the curl of Faraday's law to obtain a new vector field as:

$$\nabla \times \nabla \times \mathbf{E} = -\nabla \times \frac{\partial \mathbf{B}}{\partial t} = -\frac{\partial}{\partial t}(\nabla \times \mathbf{B}) \tag{2.9}$$

Stage 2:

Use Ampère's law to replace $\mathbf{B}$ on the right hand side of Eq. (2.9) by an expression involving $\mathbf{E}$:

$$\nabla \times \nabla \times \mathbf{E} = -\nabla \times \frac{\partial \mathbf{B}}{\partial t} = -\frac{\partial}{\partial t}(\nabla \times \mathbf{B}) = -\varepsilon\mu \frac{\partial^2 \mathbf{E}}{\partial t^2} - \mu \frac{\partial \mathbf{J}}{\partial t} \tag{2.10}$$

The right-hand side of this equation is beginning to look something like a wave equation: we have a partial second-order time derivative as required in a wave equation (and arising from the displacement current), with the time derivative of the current appearing as a source term. A problem remains however with the left-hand side. To solve this we use a standard vector identity to expand the "curl curl" operation in Stage 3.

Stage 3:

Use the vector identity (see Problem 1.7)

$$\nabla \times \nabla \times \mathbf{E} = \nabla(\nabla . \mathbf{E}) - \nabla . \nabla \mathbf{E} \tag{2.11}$$

to replace the "curl curl" operation by the difference of "grad div" and "div grad". This looks promising since we expect to obtain a "div grad" operation

29

in the wave equation, but the "grad div" term is still a problem. We can, however, use Gauss's law for the electric field to eliminate this term.

Stage 4:
Use Gauss' s law, i.e. $\nabla.E = \rho/\varepsilon$ and the assumption that $grad(\rho) = 0$, i.e. that the charge density is a constant, to set the first term in the expansion of Stage 3 to zero (if we cannot make this assumption then we still obtain a wave equation but its detailed form is much more complex than the one we are going to consider). We finally obtain the following inhomogeneous wave equation:

$$\nabla^2 E - \varepsilon\mu \frac{\partial^2 E}{\partial t^2} = \mu \frac{\partial J}{\partial t} \qquad (2.12)$$

There are four important observations arising from this equation
1. In free space, $J = 0$ but $\varepsilon = \varepsilon_0$ and $\mu = \mu_0$, so waves propagate with velocity

$$v = c = \frac{1}{\sqrt{\varepsilon_0\mu_0}} = 2.998 \times 10^8 \, m \, s^{-1}$$

2. For material with refractive index n, the velocity of propagation is modified to become, by definition,

$$v = \frac{c}{n} = \frac{1}{\sqrt{\varepsilon\mu}} = \frac{1}{\sqrt{\varepsilon_0\mu_0}} \cdot \frac{1}{\sqrt{\varepsilon_r\mu_r}} = \frac{c}{\sqrt{\varepsilon_r\mu_r}} \qquad (2.13)$$

or
$$n = \sqrt{\varepsilon_r\mu_r}$$

This is an important relation (called the "Maxwell relation") between the refractive index of a material and its electric and magnetic properties defined by the relative permittivity ε_r and permeability μ_r.
3. Radiation is caused by time-varying currents $\partial J/\partial t$. Since currents are due to moving charges, i.e. $|J| = \partial q/\partial t$ we obtain the result that radiation is due to accelerating charge.
4. When the assumption of Stage 4 that grad $\rho = 0$ is not valid, i.e. when the charge density varies in space, then we no longer have a simple wave equation. Instead we must return to Eq. (2.10) which is called a *vector wave equation* and can be written in the form

$$\nabla \times \nabla \times E + \varepsilon\mu \frac{\partial^2 E}{\partial t^2} = -\mu \frac{\partial J}{\partial t}$$

2.3 Basic mechanisms of electromagnetic wave generation

We now turn to give a more physical interpretation of this acceleration rule by considering the case of a particle with charge q moving along the x-axis (this approach is based on that developed by Tessman and Finnel; see Suggestions for further reading at the end of this chapter).

Starting at $t = 0$, the particle is at rest and then accelerates to a velocity Δv in time Δt, thereafter moving with constant velocity. We choose to observe the situation at time t where $t \gg \Delta t$. The velocity/time profile of the particle is then as shown in Figure 2.6

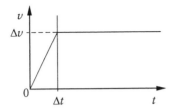

Figure 2.6 Time/velocity profile for a charged particle.

We first outline the general properties of the fields around the particle before using a geometrical argument to derive an expression for the radiated electric field.

Consider the following two limiting cases:

1. At time $t = 0$, the particle is at rest and we have an electrostatic problem. The fields around the particle are given by Gauss's law as (Fig. 2.7):

$$\mathbf{E}_1 = \frac{q}{4\pi\varepsilon_0 r^2}\,\mathbf{r} \qquad \mathbf{B}_1 = 0 \qquad (2.14)$$

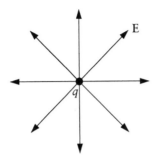

Figure 2.7 Radial electric field pattern from a point charge.

Note that the electric field is given by Coulomb's law and is radial, centred on the position of the particle at $t = 0$.

2. At time $t \gg \Delta t$ we have a case of constant current $I = q\Delta v$ and the fields are given by the equations of magnetostatics as

$$E_2 = \frac{q}{4\pi\varepsilon_0 r^2}\,r \qquad \nabla \times B_2 = \mu_0 J \qquad (2.15)$$

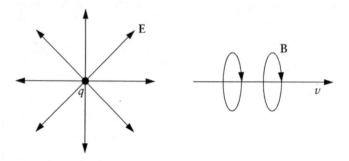

Figure 2.8 Electric and magnetic fields around a particle moving with constant velocity v.

Note that E_2 is again given by Coulomb's law, with the important difference that the field lines are now centred on the new position of the particle, which is approximately $x = \Delta vt$ (we assume $\Delta v \ll c$).

In going from case (1) to case (2), **B** has changed from B_1 to B_2. It must change during the interval Δt, and so for this short time interval we have a transient problem where the B field changes. This transient process is sustained by the coupling of **E** and **B** from Maxwell's equations as follows

Process A:
B changes from B_1 to B_2 so $\partial B/\partial t \neq 0$, causing a transient electric field E_t given by

$$\nabla \times E_t = -\frac{\partial B}{\partial t}$$

Process B:
E changes from E_1 to E_t so $\partial E/\partial t \neq 0$, which causes a change in B to

$$\nabla \times B_t = \varepsilon\mu\frac{\partial E}{\partial t}$$

This then causes another change in E given by Process A and we have a self-sustaining transient wave which propagates through space with velocity given by $1/\sqrt{(\varepsilon\mu)}$. In terms of the two processes A and B described above, we

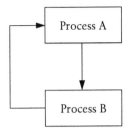

Figure 2.9 Feedback mechanism for electromagnetic wave propagation.

have the feedback wave mechanism shown in Figure 2.9.

Now we see how waves can propagate in vacuum: they need only the coupling of **E** and **B** through Maxwell's equations to sustain themselves as a propagating transient. Furthermore, we see that the effect of the presence of material is accounted for only by a change in the velocity of propagation through the parameters ε and μ. The material plays no mechanical part in the wave process as it does for sound waves; electromagnetic waves require only the sustained transient coupling of electric and magnetic fields in order to propagate.

Schematically, we can represent the wave propagation inherent in Maxwell's equations by thinking of the wave as a chain of **E** and **B** loops originating on a (changing) current element, as shown in Figure 2.10.

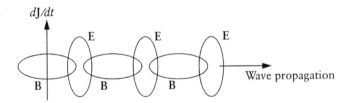

Figure 2.10 Electromagnetic wave propagation based on the curl equations.

Note that, because of the vector definition of the curl operation, these loops are in perpendicular planes, just like links in a chain.

We can quantify the radiated field from such a particle by considering the electric field lines around the particle and assuming a law of causality, which we interpret as stating that information about the movement of the particle takes a finite time to travel through space (the maximum velocity being given by c, the velocity of light in a vacuum). We then consider the electric field lines at time $t \gg \Delta t$, as shown in Figure 2.11.

The field lines form two distinct sets. In region 1 they are radial and centred on O, the position of the charge at time $t = 0$. In region 2 they are still radial but are now centred on P, the position of the charge at time t. The field

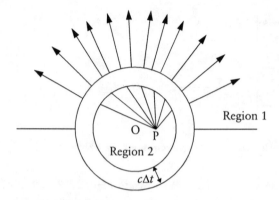

Figure 2.11 Electric field lines around and accelerating charge.

lines in region 1 (which lie outside a sphere of radius ct) cannot be centred on P since our assumption of causality prevents information about the movement of the charge reaching this zone.

Our interest now centres on the small zone between regions 1 and 2, shown as the annulus of width $c\Delta t$. (Note that the width of this zone compared to the radius ct has been greatly exaggerated for ease of viewing. In fact, we assume $t \gg \Delta t$ and so this annulus is very thin compared to the radii of the circles.)

In order to construct the field line pattern inside this annulus, we invoke Gauss's law for electric fields in free space, which states that in the absence of free charges $\mathrm{div}\,E = 0$. Geometrically this implies that field lines can only ever terminate on charges and, since there are none in free space, the field lines must connect in a one-to-one mapping across the annulus. Since the field lines on either side have different centres, this will cause a "kink" in the field pattern. Let us analyze this "kink" further by examining the path of a single field line, as shown in Figure 2.12.

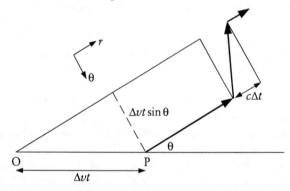

Figure 2.12 Magnification of the electric field lines around the acceleration annulus.

We see immediately that the electric field inside the annulus has both r and θ components. The θ component is transverse to r and, as well shall see, decays only as r^{-1} as opposed to r^{-2} for the radial component. At great distances from the charge this transverse or θ component will therefore be dominant. In this way Gauss's law forces the field lines to connect on a one-to-one basis, which yields a transverse or polarized wave. We are beginning to see how Maxwell's equations constrain the types of wave motion possible in electromagnetics.

To prove this property of the θ component, we use the fact that the electric field strength is proportional to the density of field lines, hence the transverse (or radial) field component is proportional to the number of field lines per unit area in the θ (or r) direction. However, we know that the radial component has a magnitude given by Coulomb's law as in Eq. (2.14).

If we can find an expression for the ratio E_θ/E_r then we can use Coulomb's law to find the transverse component E_θ. We can do this from simple geometry and noting that the field strengths are proportional to the following parameters

$$\left. \begin{array}{l} E_\theta \propto \dfrac{1}{c\Delta t} \\[2ex] E_r \propto \dfrac{1}{\Delta v t \sin\theta} \end{array} \right\} \text{ so } \frac{E_\theta}{E_r} = \frac{\Delta v t \sin\theta}{c\Delta t} \qquad (2.16)$$

By substituting for the radial field we obtain the following expression for E_θ

$$E_\theta = \frac{tq(\Delta v/\Delta t)\sin\theta}{4\pi\varepsilon_0 c r^2} \qquad (2.17)$$

but the radius of the sphere $r = ct$ and if we write the acceleration $a = \Delta v/\Delta t$ we finally obtain

$$E_\theta = \frac{qa\sin\theta}{4\pi\varepsilon_0 c^2 r} \qquad (2.18)$$

This is our main result. It gives the transverse component of the electric field inside the annulus in terms of the particle acceleration a, the angle θ and the (mean) radius of the annulus r. Note the following important features:

1. $E_\theta \propto 1/r$ and $E_r \propto 1/r^2$, so at large distances from the charge the field is purely transverse, i.e. it has no component in the radial direction. Thus electromagnetic waves are transverse waves and exhibit the phenomenon of wave polarization (see Section 2.4).
2. E_θ is proportional to acceleration as discussed above. The larger the acceleration the stronger the radiated field. This will be important when we discuss ways of designing efficient antennas for launching waves into space.

3. Note that E_θ is not uniform throughout space. We see that even for this simple case, the field is proportional to $\sin\theta$ and there is no radiation along the axis of the acceleration. If we plot this variation on a polar plot where the radius at angle θ is proportional to $\sin\theta$ then we obtain the polar plot shown in Figure 2.13.

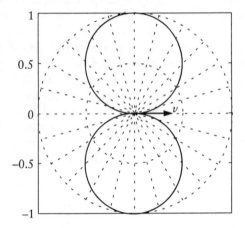

Figure 2.13 Radiation pattern for charged particle moving horizontally.

In this plot the particle is moving horizontally from left to right. Such a plot is a convenient way of displaying the radiation from antennas and is called a "radiation pattern" or "antenna polar diagram". In general, of course, it will be more complex than in this simple case and will show variations in three dimensions, not just in a plane.

This observation that radiation is dependent on spatial orientation of the source is a very important result and leads us to a second important proposition about radiated fields:

Proposition 2: Radiation only occurs from those components of acceleration **a**, that are *transverse* to the line joining the point of observation P and the source (see Fig. 2.14).

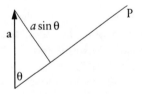

Figure 2.14 Projection of acceleration vector transverse to the direction of observation.

We see from our problem of the accelerating charge that this arises as a manifestation of Gauss's law for the electric field.

We can now use Propositions 1 and 2 to develop the impulse response of wire antenna structures. First, however, we consider the concept of wave polarization.

2.4 Wave polarization

We have seen from the above analysis that at large distances from their source, electromagnetic waves are transverse to the direction of propagation. However, this statement is not sufficient to uniquely specify the vector direction of the radiated fields since they may lie in an arbitrary direction in a plane normal to the direction of propagation and still satisfy Maxwell's equations. The various possibilites of so-called "wave polarization" can be developed using our principle of linear superposition for linear systems.

If we specify the direction of propagation as the z direction, then we can have a radiated electric field vector (or magnetic field vector) which lies along the x or y directions and still satisfies Gauss's law for the radiated field. In general, we can obtain an infinite number of solutions (all lying in the xy plane) from linear combinations of two wave functions $f(z,t)$ and $g(z,t)$ as

$$e(z,t) = \alpha f(z \pm v_x, t)\mathbf{i} + \beta g(z \pm v_y t')\mathbf{j} \qquad (2.19)$$

where α and β are real numbers and we have allowed the possibility that the velocities of the waves in the x and y directions are different. (This phenomenon of different wave velocities is called *birefringence* and occurs for light propagation in many natural crystals such as calcite and quartz.) The time $t' = t - t_0$ is used to account for the possibility that the x and y wave components are generated at different times at the source, and hence have different time origins.

The polarization of the wave is defined as the time locus of the electric field vector in the xy plane for fixed z. In general, this locus will have arbitrary form, but in practice we usually consider the following special case of functions f and g.

If we consider the case of sinusoidal wave functions, and we specify that $f = g$, we can write our general expression for $e(z,t)$ in the special form

$$e(z,t) = A\sin(\omega t \pm kz)\mathbf{i} + B\sin(\omega t \pm kz - \omega t_0)\mathbf{j} \qquad (2.20)$$

It is then easy to show that for arbitrary A, B and t_0, the locus is always an ellipse in the xy plane as shown below.

Proof that harmonic waves always generate elliptical polarizations

Choosing the plane $z = 0$ for convenience, we have

$$e_x = A\sin(\omega t)$$
$$e_y = B\sin(\omega t - \varphi) \tag{2.21}$$

In order to show that the resultant $\mathbf{e} = e_x\mathbf{i} + e_y\mathbf{j}$ is always an ellipse in the xy plane we need to first eliminate time from the expression for $|\mathbf{e}|$ and show that the resulting equation is of the general form of an ellipse given in the xy plane by the general quadratic formula

$$Ax^2 + Bxy + Cy^2 + Dx + Ey + F = 0$$
$$B^2 - 4AC < 0 \tag{2.22}$$

To do this we expand the expression for e_y using the trigonometric identity

$$\sin(\omega t - \varphi) = \sin\omega t\cos\varphi - \cos\omega t\sin\varphi$$

We can then generate the new expression

$$\left.\begin{array}{l}\dfrac{e_y}{B} = \sin\omega t\cos\varphi - \cos\omega t\sin\varphi \\[2mm] \dfrac{e_x\cos\varphi}{A} = \sin\omega t\cos\varphi\end{array}\right\} \Rightarrow \dfrac{e_x\cos\varphi}{A} - \dfrac{e_y}{B} = \cos\omega t\sin\varphi \tag{2.23}$$

To eliminate the time dependence on the right hand side, we use the result

$$e_x^{\,2} = A^2\sin^2\omega t = A^2(1 - \cos^2\omega t) \Rightarrow \cos\omega t = \sqrt{1 - \dfrac{e_x^{\,2}}{A^2}} \tag{2.24}$$

To yield, finally,

$$\dfrac{e_x^{\,2}}{A^2} + \dfrac{e_y^{\,2}}{B^2} - 2\dfrac{e_x e_y}{AB}\cos\varphi - \sin^2\varphi = 0 \tag{2.25}$$

which is the equation of an ellipse, as given in Eq. (2.22) (Fig. 2.15).

The ellipse may be specified by geometrical parameters; it has a major axis inclination angle θ defined in the range $-\pi/2 < \theta \le \pi/2$ and ellipticity angle τ such that $-\pi/4 < \tau \le \pi/4$ and

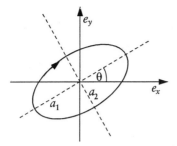

Figure 2.15 The polarization ellipse.

$$\tan \tau = \pm \frac{a_2}{a_1} \tag{2.26}$$

Note that the geometrical properties of this ellipse (its major axis inclination angle θ and ellipticity τ) are independent of time. They are defined in terms of the parameters A, B and φ as

$$\tan 2\theta = \frac{2AB\cos\varphi}{A^2 - B^2}$$

$$\sin 2\tau = \frac{2AB\sin\varphi}{A^2 + B^2} \tag{2.27}$$

An important parameter describing the shape of the ellipse is the axial ratio (AR) defined as

$$AR = \frac{a_1}{a_2} = \frac{1}{\tan\tau} \tag{2.28}$$

which is the ratio of the lengths of the major and minor axes. We shall make use of this ratio to describe the radiation from helical antennas.

Two special cases of wave polarization are particularly important:

Case 1
If $t_0 = 0$ the locus is always a straight line at an angle $\theta = \tan^{-1}(B/A)$ and the wave is linearly polarized. The axial ratio is infinite for linear polarizations. Linearly polarized waves are generated by a wide range of wire and aperture antennas, including dipoles, loops and horns.

Case 2
If $t_0 = \pm(\pi/2\omega)$ and $A = B$ we have the special case where the locus is a circle and the wave is circularly polarized. In this case the electric field vector has

the special form

$$e_L(t) = \sin(\omega t)\mathbf{i} + \sin\left(\omega t + \frac{\pi}{2}\right)\mathbf{j} = \sin(\omega t)\mathbf{i} + \cos(\omega t)\mathbf{j}$$

$$e_R(t) = \sin(\omega t)\mathbf{i} + \sin\left(\omega t - \frac{\pi}{2}\right)\mathbf{j} = \sin(\omega t)\mathbf{i} - \cos(\omega t)\mathbf{j}$$

(2.29)

Note that for circular polarization the axial ratio is unity. Note also that we can have two senses of circular polarization depending on whether the circle is traversed clockwise or anticlockwise with time. The sense of polarization is very important in determining the behaviour of a wave when it strikes a receiving antenna or scatterer.

By convention, we refer to clockwise rotation as left-sense or left-handed polarization, with anticlockwise being termed right-sense (note that the sense is defined as the direction of rotation when looking *back* towards the source of the wave). The reader should note, however, that the opposite notation is also widely used and there is ample opportunity for confusion.

We shall meet circular polarization when we consider the helical antenna (it also appears in a rather interesting form in the case of the loop antenna). In that case we shall make use of the following interesting result: we can consider a linearly polarized wave as the sum of two circular components with opposite sense. In fact any polarization can be expressed as the sum of two such circular polarizations with variable amplitudes and relative phase angle. This is easily seen from the above expressions for e_L and e_R since by simple vector addition we have

$$e_L(t) + e_R(t) = 2\sin(\omega t)\mathbf{i}$$

$$e_L(t) - e_R(t) = 2\cos(\omega t)\mathbf{j}$$

(2.30)

The polarization of an antenna is defined as the polarization of the wave it transmits. It is important to have an understanding of the polarization properties of an antenna because an efficient receiving antenna must be "matched" to the polarization of the incident wave.

Every elliptical polarization has its "orthogonal" partner for which the mismatch results in zero received signal (for linear x polarization this is simply linear y polarization, but more surprisingly perhaps, the two opposite sense circular polarizations are also orthogonal). The concept of orthogonality of two polarizations is then a mathematical statement of the mismatch between a wave and antenna polarization. Note that the condition for two polarizations to be orthogonal can be expressed in a time-independent form for harmonic waves.

If the receive antenna is matched to the wave's orthogonal polarization then the received signal will be zero, independent of other elements in the receiver system. We must bear this in mind and always consider the polarization of the radiation when analysing the time-domain response of antennas.

2.5 Impulse response of wire antennas

Antennas are metal or dielectric structures which are engineered to provide an efficient launch of electromagnetic waves into space. They are designed to exploit the acceleration rules outlined above. One of the simplest such structures (and yet one which is surprisingly efficient) is a thin straight cylindrical wire antenna of length L and wire radius r where $r \ll L$, as shown in Figure 2.16.

Figure 2.16 Cylindrical wire antenna.

We shall assume in all that follows that the wires are made of perfectly conducting material so there are no ohmic I^2R losses in the antenna. These ohmic losses are important but ignoring them does not prevent us from discussing the essential physical principles of antenna operation. Physically, we can model the metallic antenna structure as providing a sea of free electrons which can be influenced by external forces and offer zero resistance to movement. This idea will be central to our understanding of the operation of the antenna.

Consider the case where we cut such a wire in the centre and connect a voltage source between the two halves, as shown in Figure 2.17. The voltage source is assumed to be a delta function, i.e. $v(t) = \delta(t)$. Initial conditions are set so that at time $t = 0$ all currents on the wire are zero.

The voltage source is manifested as an electric field across the gap of the feed point (the relative size of which is shown greatly enlarged in Figure 2.17). The electric field has a strength given approximately by the voltage divided by the gap width. This electric field causes instantaneous charge separation at the feed point, as shown schematically in Figure 2.18.

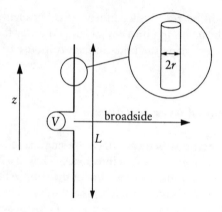

Figure 2.17 Centre-fed dipole antenna.

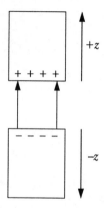

Figure 2.18 Simple model for the feed point of a dipole antenna.

This separation of positive and negative charge is accompanied by an acceleration of charge from rest. This acceleration causes radiation from the feed point. The charge movement local to the feed point then propagates as a wave along the conductors in the $\pm z$ directions.

The two charge packets move along the arms of the antenna with constant velocity (since there are no other forces on the charge). Hence there is no radiation for a time corresponding to the transit time along each arm of the antenna. If the antenna is embedded in free space and $r \ll L$ this velocity will be approximately c and so the transit time will be $L/2c$. In practice, as the wire radius increases, the velocity of propagation becomes slightly less than c and the packets take longer to reach the end of the wire. Our theory is therefore strictly valid only for wires of infinitesimal thickness (in section 4.5 we shall see how to include the finite radius of real antennas).

When the charge reaches the end it is decelerated to zero (the boundary conditions are that the current $I = 0$ at the open circuit ends of the antenna) and accelerated in the opposite direction. This causes radiation from the end-points of the antenna.

The current pulses then move along the antenna arms with constant velocity (but in the opposite direction) and no radiation. What happens next depends on the feed impedance. If the source has zero impedance (i.e. is a short circuit) then the pulses cross in the centre and no radiation occurs until they reach the ends again, where they are decelerated.

If the source impedance is not zero then we obtain some reflection of the pulses and radiation from the feed point. This process continues, with pulses propagating up and down the length of the antenna. Note that the process does not continue indefinitely (even for a perfectly conducting wire) since at each deceleration/acceleration point energy is lost to the radiation field and so the current pulse is steadily reduced in amplitude. The overall situation is depicted graphically in Figure 2.19.

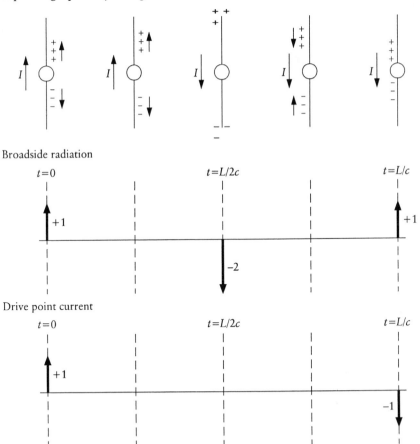

Figure 2.19 Charge movement on a linear wire antenna.

For the sake of clarity and to investigate the details of the antenna response, we assume that the source impedance is "matched" to the antenna. By "matched" we mean that we have an impedance at the source point which causes all the incident charge to come to rest (a resistor of approximately $1\,k\Omega$ is a good practical estimate of this source impedance for thin cylindrical dipoles). In this case we have the time-limited velocity/time profile for the charge pulse moving in the $+z$ direction shown in Figure 2.20.

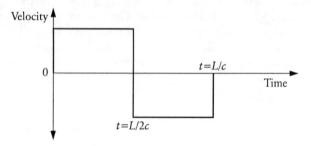

Figure 2.20 Time/velocity profile for charge moving in the +z direction.

Using our acceleration rule we can then find the contributions to the radiated field by differentiating this profile to obtain the charge acceleration shown in Figure 2.21. We see there are three delta functions: one from the source point, a negative pulse (of twice the magnitude of the first) due to deceleration/acceleration at the end of the wire, and a deceleration of the current pulse into the matched load.

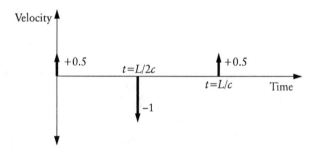

Figure 2.21 Derivative of the charge/velocity profile.

We can perform a similar analysis on the current pulse moving in the $-z$ direction to obtain the time/velocity and radiated field responses shown in Figures 2.22 and 2.23.

To find the resultant radiated field, we need to consider two more factors. First we must consider the propagation times for the various components to the observation point P, as shown in Figure 2.24. P is the observation point and we have identified the three key radiation points as 1, 2 and 3. The

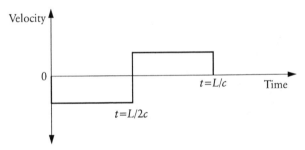

Figure 2.22 Time/velocity profile for charge moving in the −z direction.

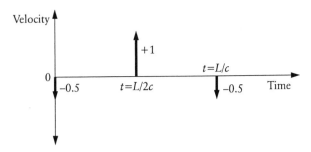

Figure 2.23 Charge acceleration for movement in the −z direction.

propagation times are r_1/c, r_2/c and r_3/c, respectively, where c is the velocity of propagation in free space. If the point P is such that $r_1 \gg L$ then we can make the following useful approximations:

$$r_1 \approx r$$

$$r_2 \approx r + \frac{L}{2}\sin\theta \qquad (2.31)$$

$$r_3 \approx r + L\sin\theta$$

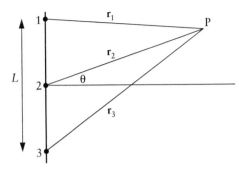

Figure 2.24 Propagation paths to the observation point P.

Secondly, in order to convert the acceleration plots into current velocity plots, we need to take account of the sign of the charge on the pulses in the $+z$ and $-z$ directions, i.e. we must make a sign correction based on the current/charge/velocity relationship:

$$I = \pm qv \qquad (2.32)$$

When we take this into account we notice an interesting result: even though the acceleration plots for the upper and lower pulse components are the negative of each other, the current derivatives (and, therefore, the radiated fields) are identical. When we combine these two features we obtain a general form for the radiated impulse response for a centre fed dipole with matched load (see Fig. 2.25), where

$$t_1 = \frac{r + \left[(L/2)\sin\theta\right]}{c}$$

$$t_2 = t_1 + \frac{L(1 - \sin\theta)}{2c}$$

$$t_3 = t_2 + \frac{L\sin\theta}{c} \qquad (2.33)$$

$$t_4 = t_1 + \frac{L}{c}$$

We see that the impulse response $h(t)$ is a sum of four delayed delta functions. As mentioned earlier, this facilitates analysis of the antenna for arbitrary excitation, since the convolution integral reduces to a weighted sum of excitation signals. Before considering the special but important case of sinusoidal excitation, we note that at broadside ($\theta = 0°$ in Fig. 2.24), the impulse response $h(t)$ has the special form shown in Figure 2.26

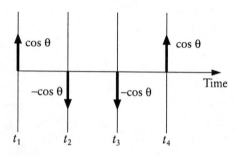

Figure 2.25 Radiated impulse response of a wire antenna of length L.

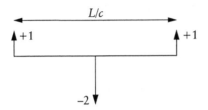

Figure 2.26 Broadside radiation from an antenna of length L.

Note that in this special case the two field pulses from the ends of the wire arrive in synchronism and yield an impulse with double the amplitude of the feed contributions. We shall see later that this causes a strong resonance at frequencies with wavelengths corresponding to important fractions of the length of the antenna.

Before we end this section we show, for comparison, the "exact" radiated field as obtained by numerical solution of Maxwell's equations for a thin wire of length 1 m, radius 1 mm and centre fed by a Gaussian pulse of FWHH = 0.5 ns and peak amplitude 10 V. The broadside radiated field at a distance of 10 m from the wire is shown in Figure 2.27.

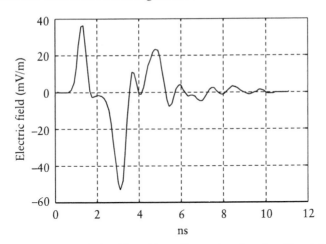

Figure 2.27 Radiated field calculated from Maxwell's equations for wire of length L = 1 m.

This field was obtained by loading the drive point with a 1 kΩ resistor to simulate matched impedance conditions. Note that the match is not perfect and we have some radiation after 6 ns, which can be attributed to multiple reflections on the wire. In the early time, however, we see that the predictions of our simple model are reasonable (at least in a qualitative sense).

We see three peaks associated with the initial drive point (which is a faithful replica of the drive voltage), the end-point reflections and current absorption in the load. Note that the polarity of the triplet is as expected, i.e. + − +, but that the relative amplitudes are slightly different from the expected 1 2 1. This is because our simple model assumes that the charge packets travel with constant velocity c along the wire, an approximation that is only rigorously justified when the wire radius tends to zero.

Nonetheless, the point should be clear that our simple acceleration model yields an acceptable first approximation to the impulse response. Only if we need to quantify the response of a real antenna must we make use of more sophisticated computer models or measurements.

2.6 Receiving antennas

Apart from their use as efficent radiators of electromagnetic waves, wire antennas are also used to receive electromagnetic signals. The main conceptual difference in this case is that the antenna is bathed in an external electromagnetic field, the influence of which generates local charge separation and hence currents in the wire. These currents can be made to flow through a load impedance and generate voltage fluctuations proportional to any modulation imposed on the incident wave. This is, of course, the basis of radio and television reception. Note that, as mentioned in Section 2.4, the wave polarization is important in determining the magnitude and direction of these induced currents. In the extreme case when the wave is orthogonally polarized to the antenna, zero currents are induced, and so no observation is made of the incident wave.

The typical situation we consider is a wire of length L with an incident plane wave field normally incident, as shown in Figure 2.28.

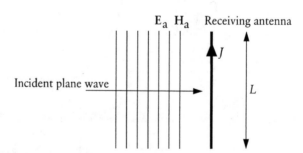

Figure 2.28 Plane wave incident on a wire antenna.

We can apply our time-domain impulse response analysis to the case of receiving antennas just as we did for transmitters. The important thing to note is that there is no fundamental reason to assume that the properties of the antenna must be the same for transmission and reception. We shall see that there are different types of distortion for radiated and received pulse signals.

In fact we shall show that there is a simple relationship between the transmitting and receiving properties of an antenna. If we know the electric field radiated by an antenna in direction θ, then we can predict the form of the current induced in the antenna by a plane wave incident from the direction θ; the induced *current* is given by the integral of the radiated *field*.

To show this, consider a charge transport picture of a wire of length L illuminated by an impulsive plane wave as shown in Figure 2.29. The charge separations give rise to charge motion and hence current flow as shown in Figure 2.29. The current flows in directions as shown, so that if we sample the current at the centre point P we obtain the current waveform shown in Figure 2.30.

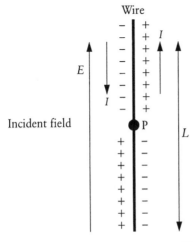

Figure 2.29   Charge movement on a wire illuminated by a plane wave.

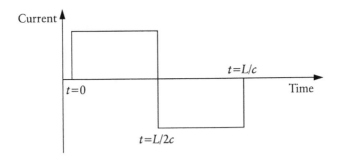

Figure 2.30   Simplified model of current flow on a wire antenna receiver.

We have assumed that the antenna is loaded with a matched impedance at the point P. The current initially rises in response to the incident impulse. This current level is maintained by charge flow for a time $L/2c$, after which the current changes direction and flows in the opposite direction for a time $L/2c$.

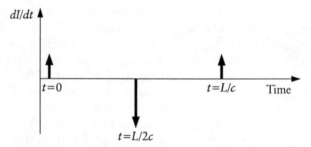

Figure 2.31 Derivative of the received current.

Thereafter the current is zero due to absorption by the matched load.

If we differentiate this current we obtain the time variation shown in Figure 2.31, which has the same form as the electric field radiated impulse response of the antenna. This result is not restricted to wire antennas. For any antenna we have the following result

> **Proposition 3:** The *current* impulse response at a point P on an antenna illuminated by an incident wave, is proportional to the integral of the radiated electric field impulse response of that antenna when driven at the same point.

To demonstrate this result we use an "exact" numerical technique (see Ch. 4) to predict the broadside field radiated by a 1 m dipole antenna with a Gaussian pulse excitation. In this case the antenna is short circuited and not terminated in a matched impedance, and we see that consequently the field contains several damped pulse contributions due to multiple reflections up and down the wire (Fig. 2.32). We can also use the numerical model to predict the current induced by a broadside plane wave with the same Gaussian

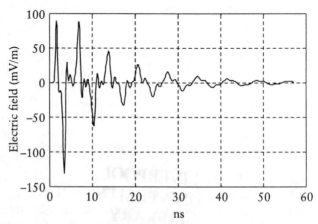

Figure 2.32 Radiated electric field at 10 m from a 1 m dipole.

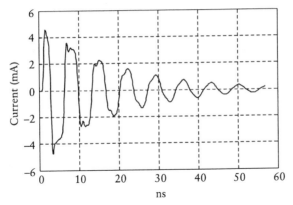

Figure 2.33 Current at the centre of a 1 m wire antenna (calculated using Maxwell's equations).

time variation. This induced current at the centre of the antenna is shown in Figure 2.33.

These two waveforms are not independent. If we differentiate the current waveform (Fig. 2.33) (using a simple numerical forward-difference approximation) we obtain the waveform shown in Figure 2.34, which has the same functional form as the radiated field shown in Figure 2.32.

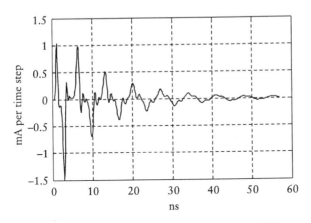

Figure 2.34 Numerical derivative of the current shown in Figure 2.33.

Hence, it can be seen that if we know the radiated *field* in any direction θ we can the easily obtain the form of the *current* induced by a plane wave incident from that same direction (and vice versa).

For sinusoidal signals this relation amounts to multiplication (for differentiation) or division (for integration) by a factor ω since

$$\frac{\partial}{\partial t}\exp(j\omega t) \equiv j\omega \exp(j\omega t)$$

51

As a result, the variation of radiation with θ (or radiation pattern) for sinusoidal excitation of fixed frequency has the same form for the transmit and receive modes. As we have seen above, this is not the case for pulse excitations, where considerable differences in waveform shape occur between transmitted and received waveforms (Figs 2.33 & 2.34).

Suggestions for further reading

The following books provide a good account of the basic mechanisms of electromagnetic radiation and discuss important properties such as radiation patterns and polarization. However, they tend to emphasize the sinusoidal response approach. Again they are recommended as much for their diversity of approach as for their clarity of exposition.

Born, N. & E. Wolf 1989. *Principles of optics.* Oxford: Pergamon Press.
Hecht, E. & A. Zajac. *Optics.* New York: Addison-Wesley.
Jackson, J. D. 1975. *Classical electrodynamics.* New York: Wiley.
Kraus. J. D. *Antennas.* New York: McGraw Hill.
Mott, H. 1992. *Antennas for radar and communications: a polarimetric approach.* New York: Wiley.
Papas, C. H. 1988. *Theory of electromagnetic wave propagation.* New York: Dover Press.
Read, F. H. 1980. *Electromagnetic radiation.* New York: Wiley.

The following specialist articles are recommended for the details they provide on the time-domain approach to radiation.

Bennet, C. L. & G. F. Ross 1978. Time-domain EM and its applications. *Proceedings of the IEEE* **66**, 299–318.
Felsen, L. B. (ed.) 1976. Transient EM fields. *Topics in applied physics* vol. 10. Berlin: Springer Verlag.
Landt, J. A. & E. K. Miller 1976. Transient response of an infinite cylindrical antenna and scatterer. *IEEE Transactions* **AP-25**, 246–50.
Miller, E. K. & J. Landt 1980. Direct time-domain techniques for transient radiation and scattering from wire antennas. *Proceedings of the IEEE* **68**, 1396–1423.
Tessman, J. R. & J. T. Finnel 1967. Electric field of an accelerating charge. *American Journal of Physics* **35**, 523.

Problems

2.1   A particle with electrical charge q is moving with simple harmonic motion such that its position vector is given by $r = x_0 \sin(\omega t)\,\mathbf{i}$.

Calculate an expression for the electric field radiated by this particle in the yz plane $x = 0$.

2.2 If the particle motion now becomes circular such that its position vector is $r = x_0 \sin(\omega t)\,\mathbf{i} + y_0 \cos(\omega t)\,\mathbf{j}$, calculate the radiated field in the following directions

    i) $\mathbf{r}_p = \mathbf{k}$
    ii) $\mathbf{r}_p = \mathbf{i}$
    iii) $\mathbf{r}_p = \mathbf{i} + \mathbf{j}$

2.3 A wave travels in the z direction with electric field given by

$$e = 2\sin(\omega t - kz)\,\mathbf{i} + \sin\!\left(\omega t - kz - \frac{\pi}{6}\right)\mathbf{j}$$

Find the major axis inclination angle of the polarization ellipse. Also calculate the axial ratio and the rotation sense of the wave.

2.4 Write the following field components in terms of the sum of two circularly polarized wave components of opposite sense.

    i) $e = \cos(\omega t - kz)\,\mathbf{i}$

    ii) $e = 2\cos(\omega t - kz)\,\mathbf{i} + \cos\!\left(\omega t - kz - \frac{\pi}{6}\right)\mathbf{j}$

    iii) $e = \cos\!\left(\omega t - kz + \frac{\pi}{4}\right)\mathbf{i} + \cos\!\left(\omega t - kz - \frac{\pi}{4}\right)\mathbf{j}$

2.5 Find an approximation based on delta functions for the radiated field impulse response of a matched linear wire antenna of length L when the feed point divides the antenna in the ratio $m:n$ and $m > n$. Hence sketch an estimate for the antenna radiation impulse response from a 2 m dipole, when the feed point divides the antenna in the ration $3:2$.

2.6 A matched wire antenna of length $L=4$ m is fed with a Gaussian pulse with FWHH $= 500$ ps. Estimate the form of the radiated field at broadside and at $60°$ from broadside if the antenna is centre fed. Sketch the form of the broadside radiation if the pulse length is changed so that FWHH $= 500$ ns.

2.7 The same antenna as described in question 2.6 is now used to receive pulse signals from a direction $\theta = 30°$. Estimate the form of the impulse response current at the terminals of the antenna.

Chapter 3

Antenna resonance and harmonic excitation

3.1 Resonance in antenna systems

Whereas there are several important applications where direct pulse excitation of antennas is used (for example, in radar and broad-band S-matrix measurements of microwave circuits), by far the most common form of excitation of wire antennas is by a single harmonic carrier at frequency f_c.

Applications in television, radio broadcasting and portable telephony all use sinusoidal excitation of the antenna, together with narrow band modulation to carry the voice/picture information. It is therefore of great practical interest to consider antenna voltage excitation functions of the form $x(t) = \sin(\omega_c t + \phi)$.

Remember that our objective is to engineer structures which radiate efficiently, i.e. as large a fraction as possible of the source energy must be transferred to the radiated field. There are three factors to be considered when trying to achieve this high efficiency:

- We need to design an antenna where the charge acceleration is maximized to obtain a large radiated field.
- We need to ensure that the various components of acceleration propagate to the desired points in space and add constructively to give a large resultant field.
- We need to ensure that the voltage or current source generator is matched to the antenna, i.e. we need to consider the impedance of the antenna as seen from the source.

As mentioned earlier, we can obtain the radiated field at any angle θ by convolution of the impulse response in that direction with a sinusoid. Before we carry this out analytically, it is instructive to consider a pictorial interpretation of this convolution process.

Convolution involves reflection of our impulse response about the origin followed by a slide, multiplication and summation process. To illustrate, let us convolve an input function of the form of a rectangular pulse $x(t)$ with an impulse response which consists of a pair of delta functions shown as $h(t)$ in Fig. 3.1.

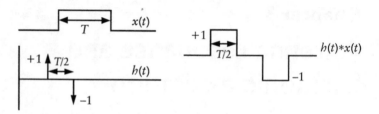

Figure 3.1 Convolution of a doublet with a rectangular pulse.

We see that the convolution process has distorted the input function $x(t)$ to produce an output function which is a rough approximation of the derivative of the input. Such an impulse response, which consists of a pair of delta functions of opposite sign, in the limit as their separation tends to zero, is called a *doublet* (denoted by D2) and may in a formal sense be considered to be the derivative of a delta function. Similarly, we may define a triplet D3 as the second derivative of the delta function (see Fig. 3.2). We have seen that such functions arise naturally in the description of wire antennas (Chapter 2).

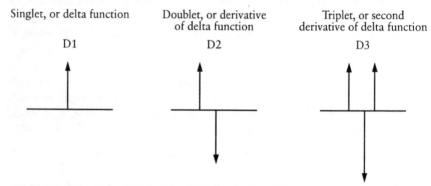

Figure 3.2 The singlet (D1), doublet (D2) and triplet (D3) functions.

The effect of convolving a doublet D2 of width τ with a function $f(t)$ is to produce a function which in the limit as τ tends to zero is proportional to the derivative of the function since

$$\int \left[\delta(t - t') - \delta(t - t' + \tau) \right] f(t) \, dt = f(t') - f(t' - \tau) \tag{3.1}$$

and since τ is small we have

$$f(t' - \tau) \approx f(t') - \frac{df}{dt} \tau \tag{3.2}$$

and so we have the result

$$f * D2 \approx \frac{df}{dt} \tau \tag{3.3}$$

We shall use this result in Chapter 5 where we will consider the radiation from aperture antennas.

We now consider the case when $x(t)$ is a sinusoidal function and $h(t)$ our triplet of delta functions for broadside radiation from a dipole antenna (as discussed in Chapter 2). The first point to note is that, unlike the example considered in Figure 3.1, where pulse distortion occurs, the radiated field (or output) will in this case always be sinusoidal. This is due to the fact that the resultant field is a linear combination of sinusoids of the same frequency as the excitation. It is a rather special property of sinusoids that any such linear sum can always be expressed as a single sinusoidal component.

We can investigate the effects of changing the carrier frequency f_c by considering the convolution of our impulse response with sinusoids of constant amplitude, but varying period, as shown schematically in Figure 3.3.

In Figure 3.3a we show the limiting case when the period of the excitation is much longer than the duration of the antenna impulse response (for simplicity we consider the case $\theta = 0°$, i.e. broadside radiation). In this case we see that as the triplet $h(t)$ slides across the sinusoid, it yields a very small resultant field. In fact, in the limit as the period tends to infinity, i.e. as the frequency of the excitation tends to zero, we see that the response will always be of the form $1 - 2 + 1 = 0$. This is a very important result and explains why it is difficult to radiate energy at low frequencies: it is simply due to cancellation of the various acceleration components radiating from the feed and ends of the wire.

We can quantify "low frequency" in this context since we know that the duration of the impulse response is L/c. Low frequencies are those with corresponding wavelengths ($\lambda = c/f$) which are large compared to the antenna length L. For example, the radiation of 50 Hz power signals (with a corresponding wavelength of $\lambda = 6000$ km) from electrical power lines of even 100 km length is very inefficient.

In fact, from our earlier discussion we see that a wire which is short compared to a wavelength (which we call an "electrically short antenna" to distinguish the fact that its physical length may still be large) radiates the second derivative of the excitation voltage. In the case of a sinusoid this second derivative is again a sinusoidal function, but for pulse excitation there will be some pulse distortion. We shall see in Chapter 5 that small aperture antennas such as the transverse electromagnetic (TEM) horn also radiate the second derivative of the excitation signal. In this sense the second-derivative radiation law is a generic property of electrically small antennas.

In Figure 3.3b we have the case where the length L of the antenna corresponds to half a wavelength of the excitation. In this case, as we slide the triplet across the sine wave we see that the cancellation which occurred for large λ is no longer a problem. However, this is not the most efficient case (we shall see later that for this length, the antenna has important impedance properties).

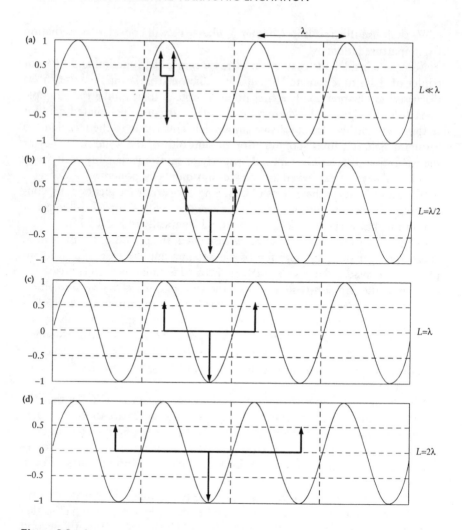

Figure 3.3 Antenna resonance.

In Figure 3.3c we show how resonance can occur when the antenna is one wavelength long. Here we see that when the +1 −2 +1 triplet multiplies the sinusoid we obtain a situation where the three components always add constructively to yield a maximum in the radiated field. Unfortunately, we shall see that this situation also corresponds to a maximum in the input impedance (which typically has a value of a few thousand ohms). This places a practical limitation on the use of full-wavelength antennas, since it is difficult to feed high input impedance antennas from commonly used sources and transmission lines (which typically have impedances from a few up to a few hundred Ohms).

Finally, in Figure 3.3d we show the case when the antenna length is twice the wavelength of the excitation. Here we see that instead of constructive addition of the triplet components, they always add destructively to obtain a radiation null. Note that this minimum occurs only in the broadside direction and the antenna will still radiate at other angles θ.

We can quantify these observations by considering the formal convolution of our triplet with the sinusoid to obtain the radiated field at broadside as

$$y(t) = \sin(\omega t) - 2\sin\left(\omega t + \frac{\omega}{c}\frac{L}{2}\right) + \sin\left(\omega t + \frac{\omega}{c}L\right) \tag{3.4}$$

A similar expression can be obtained for the radiation at angle θ by using the full impulse response discussed earlier (Eq. (2.33)).

The first point to note is that we have a weighted sum of sinusoidal components with different amplitudes and phases. The phase terms can be more conveniently expressed in terms of the wave number of the source, k, defined as $k = \omega/c = 2\pi/\lambda$.

We see that the phase shift is then expressed in terms of the product kL as

$$y(t) = \sin(\omega t) - 2\sin\left(\omega t + \frac{kL}{2}\right) + \sin\left(\omega t + kL\right) \tag{3.5}$$

and that this product is important in determining the efficiency of the radiating system. We can obtain the modulus of this expression as a function of kL by forming the sum of complex phasors corresponding to Eq. (3.4).

$$y(\alpha) = 1 - 2\exp(i\alpha/2) + \exp(i\alpha) \tag{3.6}$$

where $\alpha = kL$. The modulus of this complex function is given by

$$|y(\alpha)| = \sqrt{6 + 2\cos\alpha - 8\cos\frac{\alpha}{2}} \tag{3.7}$$

This modulus is plotted in Figure 3.4 for $0 < \alpha < 4\pi$, corresponding to $0 < L/\lambda < 2$, (note that the plot is periodic with period 4π).

Note how all the features discussed in relation to Figure 3.3 are confirmed, namely:

1. As L/λ tends to zero, the radiated field tends to zero.
2. The radiated field reaches a maximum at $L/\lambda = 1$ and then falls to zero when $L/\lambda = 2$. As we shall see, impedance limitations prevent us from exploiting the resonance at $L = \lambda$ in practical antennas.
3. At $L/\lambda = 0.5$ there is no maximum in the radiated field. However, this value has special practical significance, since for this antenna length the input impedance has a minimum value.

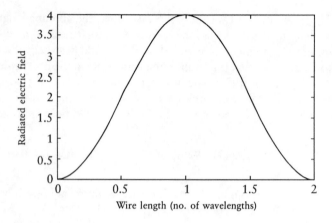

Figure 3.4 Broadside wire radiation as a function of frequency.

In practice, the antenna is seldom terminated in a matched load at the feed point and a more realistic situation is when the feed point has a low impedance or represents a short circuit. In this case we obtain more terms in the impulse response as the current waves reflect up and down the wire. We shall now show that this has the effect of sharpening the resonance structure of the sinusoidal response without changing the position of the resonant frequencies.

We consider a simplified model of the radiated field from a centre-fed dipole, as shown in Figure 3.5. Here we model the response as a finite sum of n delta functions. The amplitude of each component is attenuated by a constant factor $|r| < 1$ and the spacing is constant at τ seconds (which is related to the transit time along the dipole).

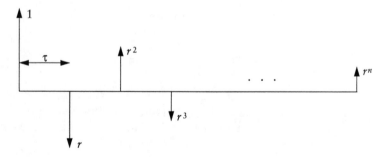

Figure 3.5 Model of the radiated impulse response of a short-circuited dipole antenna.

The impulse response then has the form

$$h(t) = \sum_{m=0}^{n} r^{m}\,\delta(t - m\tau) \tag{3.8}$$

If we now convolve this series with a sinusoid of frequency ω we obtain the phasor sum

$$H(\omega) = 1 + z + z^2 + z^3 + \ldots + z^{n-1}$$

$$z = r\exp(i\alpha) \qquad\qquad |r| < 1 \qquad\qquad (3.9)$$

$$\alpha = \omega\tau$$

This series can be summed to obtain

$$|H(\omega)| = \left|\frac{1-z^n}{1-z}\right| \qquad\qquad (3.10)$$

Figure 3.6 shows a plot of this function for a range of values of n, i.e. for increasing numbers of terms in the impulse response and for fixed $r = -0.9$. We see that as the number of terms increases so the resonance becomes higher and narrower, indicating that the antenna is becoming more efficient at radiating energy but at the price of narrower bandwidth. We shall see that this trade-off between efficiency and bandwidth is fundamental in the design of antennas. As noted above, the narrowness of the bandwidth is related to the number of delta functions in the antenna impulse response.

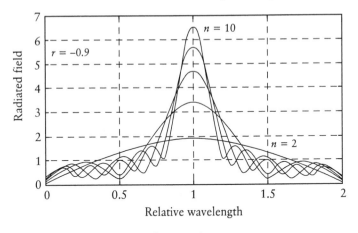

Figure 3.6 Resonance of an antenna as a function of n.

3.2 Input impedance of wire antennas

Just as we used the physics of radiation mechanisms to obtain an approximation to the radiated field from a wire antenna driven by a delta function, we can use a similar argument to obtain an estimate for the feed point current.

Armed with such an estimate for the impulse response, we can then obtain the current for sinusoidal excitation using a convolution integral.

Impedance is a concept defined for sinusoidal excitation of linear electrical systems and is defined as a function of frequency $Z(\omega)$ as the ratio of voltage and current such that

$$Z(\omega) = \frac{V(\omega)}{I(\omega)} \tag{3.11}$$

We can estimate this function by first generating the impulse response for the feed current and then using a convolution sum to generate the current at frequency ω.

Note that the voltage function is generated by convolving a delta function (the drive signal) with a sinusoidal function, which yields a complex phasor with unit magnitude, independent of frequency. Therefore, our estimate of impedance at frequency ω is simply related to the reciprocal of the observed current at frequency ω.

From our discussion of the current pulse propagating up and down the wire (Fig. 2.19) we can see that the feed-point current impulse response is a pair of delta functions separated by the wire transit time, as shown in Figure 3.7.

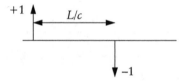

Figure 3.7 Simple model of the input current at the feed point of dipole antenna.

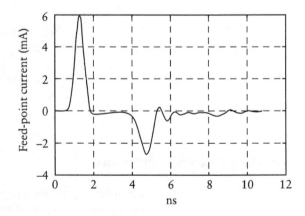

Figure 3.8 Feed-point current obtained from Maxwell's equations.

For comparison, Figure 3.8 shows the "exact" current at the feed point, as predicted by a full numerical solution of Maxwell's equations for a thin wire of length 1 m and radius 1 mm fed by a Gaussian pulse with a full width at half height (FWHH) of 0.5 ns and 10 V peak amplitude. Again we have point loaded the antenna with a 1 kΩ resistor to simulate matched impedance conditions, and again we notice that the match is not quite perfect, with some late-time current due to multiple reflections.

The polarity of the second pulse is correct, but its amplitude is less than expected from our acceleration model. This is due to antenna radiation losses not accounted for in our simple model. Nonetheless we see that our simple impulse approximation provides a good estimate and will permit us to discuss general features of the impedance properties of wire antennas. For this reason, we keep the simple approximation in the following.

From this we can obtain the feed current for sinusoidal excitation by convolution as

$$i(t) = \sin(\omega t) - \sin(\omega t + kL) \tag{3.12}$$

We can obtain the modulus of the feed current by forming the expression:

$$|i(\alpha)| = |1 - \exp(i\alpha)| = \sqrt{2(1 - \cos\alpha)} \tag{3.13}$$

Figure 3.9 shows the variation of the relative feed current for the range $0 < L/\lambda < 2$. Comparing this with Figure 3.4 we notice several important features:

1. As the electrical length tends to zero, the feed current is given by the derivative of the input voltage and tends to zero. Thus the input impedance (which is inversely proportional to the feed current) tends to infinity. This fits our expectation of the input impedance of a piece of

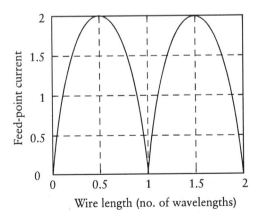

Figure 3.9 Variation of feed current with frequency.

wire which is open circuited (i.e. not connected in any closed circuit). We can see, however, that this situation changes as the frequency is increased, when the input impedance falls with frequency.

2. The input current reaches a maximum when $L/\lambda = 0.5$, i.e. when the antenna is half a wavelength long. The input impedance reaches a minimum at this point. We shall see later that it achieves a value of $73\,\Omega$ for a wire antenna in free space. This low value makes it easier to drive current onto the wire than in the case when $L/\lambda = 1$, and is more easily matched to typical transmission lines.

3. The case $L/\lambda = 1$ leads to zero feed-point current and so to a maximum in the input impedance (it is infinite according to our simple model, but in practice it rises to a few thousand Ohms). We saw earlier that this value generates a maximum in the radiated field but, since it is so difficult to drive current onto the antenna, we prefer in practice to use the overall combination of low impedance and moderate radiated field that occurs for a length half a wavelength.

We have seen that when designing antennas we need to balance two effects. The first is to maximize the charge acceleration on the antenna and to make sure that the various acceleration components add constructively in the far field. However, a second equally important effect is to match the impedance of the feed point to that of the generator/transmission line.

We have seen how by using a time-domain approach and simple approximations to the impulse response we can explain many of the qualitative features of the operation of dipole antennas. We now apply the same methods to discuss a class of antennas which employ travelling waves, which can be used over a broader range of frequencies than can the simple dipole antenna discussed so far.

3.3 Broad band wire antennas

We have seen that one of the problems of designing efficient radiating systems is that they tend to be narrow band, i.e. they operate efficiently only over a narrow range of frequencies centred on a wavelength corresponding to some fraction of the dimension of the antenna. This narrow band restriction can be traced back to the form of the impulse response.

An ideal broad band antenna has a radiated field impulse response consisting of a single delta function. Such an antenna would deliver equal power at all frequencies and would produce no distortion of pulse excitations. We have seen, however, that reflections at the boundaries of the antenna yield impulse waveforms which consist of a sum of delayed delta functions. One clear objective to obtain broad band antennas is then to modify the antenna to limit the impulse response to as short a time as possible. Indeed, we have already employed such a method when discussing the impulse response of the dipole in Chapter 2 where we assumed that the feed point was in some way "matched" to the antenna so that we limited our impulse response to a total duration of L/c seconds.

In this section we investigate how to generalize this principle to obtain antennas which operate over a wider range of frequencies than the simple dipole. We can make use of four simple principles to achieve this objective:

Resistive loading

We can place resistive point loading at the end or feed points of the antenna to absorb the current pulses excited on the antenna. Such resistively loaded dipoles tend to be inefficient (since much of the pulse energy is absorbed in the load resistors) and so are limited to low-power applications. We can improve on simple point loading by providing distributed resistive loading along the length of the antenna. When designing such antennas, we want the profile of resistivity to increase away from the feed point so as to provide a good match and yet reduce the reflections for the ends of the wire. Such resistively loaded dipole antennas are widely used for broad band applications such as ground probing radar (GPR) where short pulses are transmitted into the ground and used to detect buried objects such as water pipes and underground cavities.

Electrically large antennas

As an alternative approach, we can make the antenna larger and so push the late-time reflections further out in time. They will then limit only the very low frequency operation of the antenna. A limiting case of this would be to employ an infinitely long dipole. This would have an ideal impulse response and act as a broad band antenna. Long wires have been used for just this purpose, although as we shall see there are often better, more compact options. Furthermore, this approach only works if the current pulse is non-radiating as it moves along the antenna. Although we have assumed our wire dipole does not radiate while the pulse travels along its length, such an approximation is not strictly valid: it turns out that such radiation-free propagation can only occur when the antenna matches some special co-ordinate systems for which the Maxwell equations have a solution in terms of non-radiating current and voltage pulses. An example of this type is the biconic antenna which has a cross-section of the form shown in Figure 3.10.

If we could make such an antenna infinitely long then it would have an ideal impulse response. In practice, of course, we must truncate the design to some length L and so we obtain reflections of the current pulse from the ends of the antenna, just as we did for the dipole. This will limit the useful bandwidth. Nonetheless, such antennas have good radiation and impedance properties over a wide range of frequencies and

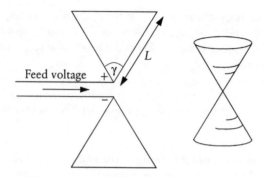

Figure 3.10 The biconic antenna.

are widely used for broad band measurement of electromagnetic radiation from electrical equipment for electromagnetic compatibility (EMC) testing.

The log-periodic antenna

We can insist that the antenna operates over a wide frequency band but with only a narrow *instantaneous* bandwidth (i.e. at any particular instant we feed the antenna with a narrow band signal). In this way we can design antennas which distort pulse excitation but can radiate a range of narrow band signals very efficiently. Log-periodic arrays are an important example of this type of antenna.

The basic idea of log-periodic antennas is to provide a set of elements with different lengths and hence resonant frequencies so that over a band of frequencies one or other of the elements will always be in resonance. The price paid for this is that the elements are spatially separated and so the time origin for the impulse response moves, depending on which element is in resonance. This is a cause of distortion if such antennas are used for pulse transmission. The basic structure of a log-periodic array is shown in Figure 3.11.

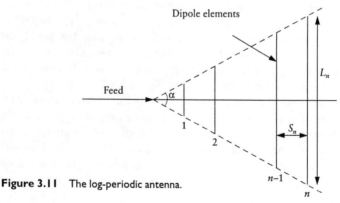

Figure 3.11 The log-periodic antenna.

At lower frequencies the dipoles at the front of the array are in half-wave resonance while those at the rear are small and so radiate very little energy. As the frequency increases the dipoles nearer the centre come into resonance. In this way the antenna operates over a wider range of frequencies than does a single element.

In order to achieve this behaviour, the array has variable spacing S_n and antenna lengths L_n such that the included angle α is a constant. This may be defined from the constant ratio β as

$$\frac{L_{n+1}}{L_n} = \frac{S_{n+1}}{S_n} = \beta$$

$$\frac{L_n}{L_1} = \beta^n = F$$

(3.14)

where F is the frequency ratio of the bandwidth of the array. For example if we choose $\beta = 1.19$ and $n = 4$, then $F = 2$, i.e. the array operates efficiently over a 2 : 1 bandwidth.

Curved structure antennas

We can modify the shape of the antenna to make the energy loss to radiation more gradual than the sudden acceleration caused by the endpoints of the dipole. This may be termed "acceleration engineering" where we make the currents flow in paths which cause efficient radiation over a broader band of frequencies. Examples such as the helix and loop antenna provide important examples of this class.

We now consider some of these options in more detail, beginning with the idea of using resistive loading to limit the impulse response duration.

One simple method for achieving broad band operation of a wire antenna is to place 'matched' resistive point loading at the end or feed points of the antenna. In the example shown in Figure 3.12 we feed the antenna at one end to obtain a travelling wave in the $+z$ direction only. Such feed systems are widely used for horizontal wire antennas at low frequencies, where the excessive physical length of the antenna precludes vertical operation.

Since this class of antenna involves waves propagating in one direction only, they are termed "travelling wave systems" and, as we shall see, they are more broad band than the dipole antenna, which involves waves travelling in the $\pm z$ directions and for sinusoidal excitation results in resonant or standing waves.

Using the same techniques employed for analysing the dipole, the impulse response for the radiated electric field from such a travelling wave antenna can be shown to consist of a pair of delta functions, one from the feed point and the other from the terminating load, with relative timing as shown in Figure 3.13.

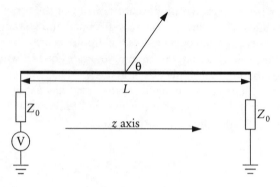

Figure 3.12 The horizontal travelling wave antenna.

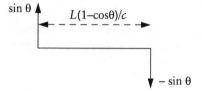

Figure 3.13 Radiated impulse response of travelling wave antenna.

From this we see that the radiated field shows some resonant behaviour with a broadside sinusoidal response of the form:

$$y(t) = \sin(\omega t) - \sin(\omega t + kL) \tag{3.15}$$

The magnitude of this function is similar to the kL variation of feed-point *current* for the centre-fed dipole (see Fig. 3.9) with a maximum when $L = 0.5\lambda$ and a minimum at $L = \lambda$. However, since the matching resistors are in place, there will be no resonance sharpening as observed in Figure 3.6.

The feed-point current impulse response is more interesting; it has the form of a single delta function (assuming perfect matching through the resistors Z_0), showing that the input impedance does not vary with frequency. The main advantage of such an antenna is this broad band feeding capability. Note that in practice we need to obtain the value of Z_0, a matched impedance for the antenna. In chapter 4 we will see how to estimate such a value but for the moment we note that there will always be power dissipation in these terminating resistors and so the antenna will be less efficient than the centre-fed dipole. This is a fundamental trade-off in antenna design; we can generally achieve broad band operation or high efficiency, but not both.

We now turn to consider a second approach for designing efficient broad band antennas: namely to shape the antenna structure so as to force the current into curved paths which cause charge acceleration and so radiation across the whole antenna structure. Two examples of this approach are the loop and helical antennas.

3.4 The circular loop antenna

The loop antenna offers a very different design topology to the dipole, and as such has very different radiation and impedance properties. Historically, a rectangular loop antenna was used by Heinrich Hertz in the very first demonstration of radio wave propagation. Hertz's basic arrangement is shown in Figure 3.14. He used a loaded dipole antenna as a transmitter, fed by an inductive spark gap generator. To receive the waves radiated by this antenna he used a loop of the dimensions shown. In this section we investigate the properties of such loop antennas.

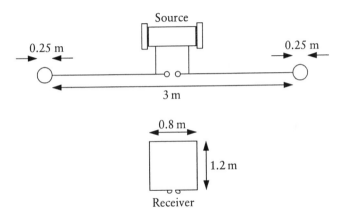

Figure 3.14 Hertz's radio system of 1886.

We consider the problem of finding the radiated electric field impulse response of a circular loop antenna, as shown in Figure 3.15.

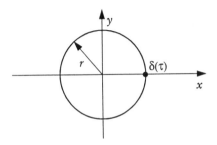

Figure 3.15 Circular loop antenna.

The delta function drive generates radiation from the feed point, just as in the case of the dipole antenna, and then two counter-propagating current pulses move around the circle of radius r. Again, as with the dipole, we shall assume that the source point is matched, so after one circuit of the antenna the pulses are decelerated to rest, causing a negative delta function radiation contribution from the feed point. If the load point is a short circuit then the current waves will travel around the loop several times before decaying to zero because of dissipation to the radiation field. In this case the resonance of the antenna will be sharper than for the single transit. For simplicity, however, we consider only the former case.

In between the two delta functions, we assume the current pulses move with constant speed (assumed to be c) but their velocity vectors change continuously. Hence, unlike the dipole, there is radiation from each point of the loop antenna.

We can analyse this continuous radiation component by considering the clockwise and counter clockwise pulses separately. There are three stages to the analysis of such antennas:

Charge acceleration model

The counter clockwise pulse has position, velocity and acceleration vectors given by

$$
\left.
\begin{aligned}
\mathbf{r} &= r\cos\beta t\, \mathbf{i} + r\sin\beta t\, \mathbf{j} \\
\frac{d\mathbf{r}}{dt} &= r\beta(-\sin\beta t\, \mathbf{i} + \cos\beta t\, \mathbf{j}) \\
\frac{d^2\mathbf{r}}{dt^t} &= r\beta^2(-\cos\beta t\, \mathbf{i} - \sin\beta t\, \mathbf{j})
\end{aligned}
\right\}
\quad \mathbf{a}_R = qr\beta^2(-\cos\beta t\, \mathbf{i} - \sin\beta t\, \mathbf{j}) \quad (3.16)
$$

where a is the current velocity which causes radiation, and i and j are unit vectors in the x and y directions. From our discussion of wave polarization (Chapter 2), we see that this pulse generates a right-hand circularly polarized wave with amplitude dependent on the loop radius r and angular frequency β. The factor β can be easily determined since we assume the pulse propagates around the loop with constant speed c, so for a loop of radius r the angular speed is just

$$
\beta = 2\pi(2\pi r/c)^{-1} = \frac{c}{r} \ \text{rad}\,\text{s}^{-1} \quad (3.17)
$$

and the acceleration is proportional to $\beta^2 r = c^2/r$.

Likewise, the clockwise pulse has position and acceleration vectors given by

$$\left.\begin{array}{l} \mathbf{r} = r\cos\beta t\,\mathbf{i} - r\sin\beta t\,\mathbf{j} \\ \dfrac{d^2\mathbf{r}}{dt^2} = r\beta^2(-\cos\beta t\,\mathbf{i} + \sin\beta t\,\mathbf{j}) \end{array}\right\} \quad \mathbf{a}_L = -qr\beta^2(-\cos\beta t\,\mathbf{i} + \sin\beta t\,\mathbf{j}) \quad (3.18)$$

This pulse component generates a left-hand circularly polarized wave.

Propagation of radiation to the observation point P

We now calculate the resultant field along the z axis of the antenna (Fig. 3.16). The total field will depend on the sum of the acceleration vectors for each pulse. (Note that there are no propagation distortions since the distance from the radiation points on the loop to an observation point on axis are all equal.)

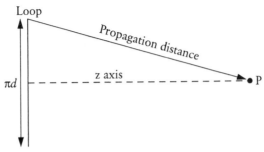

Figure 3.16 Propagation of radiation to observation point P.

The total radiated field at time t can be found by integration of the acceleration vector around the loop. If we consider just one of the sinusoidal components of this acceleration then we obtain an integral of the form

$$f(t) = \int_0^{\frac{2\pi r}{c}} \delta(t - t_0)\cos\beta t_0\, dt_0 = \cos\beta t\,\Big|_{t=0}^{t=\frac{2\pi r}{c}} \quad (3.19)$$

where $t_0 = s/c$ and s is the distance around the loop. We can interpret this by saying that at any time t we find the field $f(t)$ by adding all contributions from around the loop, but only the contribution for which $t = t_0$ contributes a non-zero value, so the net radiated field is a single cycle of a cosinusoidal pulse.

When we integrate all four components of acceleration and perform vector addition, the result is the following expression for the radiated field along the z axis:

$$\mathbf{a}_L + \mathbf{a}_R = -2r\beta^2 q\sin\beta t\,\mathbf{j} \quad (3.20)$$

where we see that the two opposite-handed circular polarizations combine to yield a linearly polarized field. Hence, despite the fact that loops constrain the currents to flow in circular paths, they radiate linearly polarized waves.

The radiated impulse response consists of a pair of delta functions plus a single cycle of a sinusoid, as shown schematically in Figure 3.17. Note that the field is linearly polarized in the y direction.

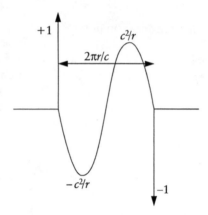

Figure 3.17 Radiated impulse response of a circular loop antenna.

Figure 3.18 shows the "exact" field obtained by numerical solution of ME. The example shown is for a circular loop of radius 1 m fed by a Gaussian pulse with FWHH = 2 ns and 10 V peak amplitude. The field is calculated for a point on axis a distance 10 m away from the loop. Note that there is an initial pulse from the drive point followed by a single-cycle sine wave with polarity as expected from above. The second pulse from the termination impedance (set to 1 kΩ in this example) is wider than the initial pulse due to the loss of high fre-

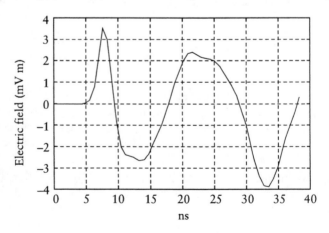

Figure 3.18 Calculated radiation from a circular loop antenna of radius r = 1 m.

quencies to radiation as the pulse propagates around the loop. Note that the single cycle sinusoid has a period of approximately 21 ns as expected from the time taken for light to travel once around the circumference.

Impedance of the loop antenna

The feed current impulse response has the form of a pair of delta functions as shown in Figure 3.19, where we have accounted for radiation losses around the loop by attenuating the second pulse to an amplitude a (which typically has a value of $a = 0.5$). Figure 3.20 shows the "exact" feed current for a 1 m radius circular loop fed with a Gaussian pulse of 2 ns FWHH and 10 V peak amplitude (the same example as used in Fig. 3.18 for the field calculation). The feed point has a source impedance of 1 kΩ.

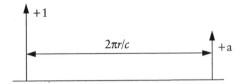

Figure 3.19   Feed current model for a loop antenna.

We see that the current is limited to two pulses separated in time by approximately 21 ns, which corresponds to the transit time for light around the loop.

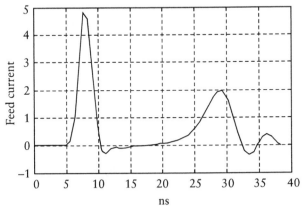

Figure 3.20   Calculated feed-point current for a loop.

We see that there are several important differences between the loop and dipole antenna impulse response waveforms. In particular, we note the following:

- The two delta functions in the radiated field response from the drive point acceleration and deceleration have opposite signs (in contrast to the dipole drive point contributions). This is due to the fact that the current travels around the antenna and does not suffer reflections as it did for the dipole. This behaviour is confirmed by the drive point current response which consists of two delta functions of the same polarity.
- The radiated field for the loop contains an extra contribution from the transit time around the loop. This works in opposition to the drive point acceleration and deceleration. For this reason, electrically small loops are very poor radiating systems. For any given size, a dipole is a much more efficient transmitter. Consequently, loops are seldom used as transmitters, although they find widespread use as receiving antennas.

We can analyse the contributions of the two parts of the radiated impulse response as follows. The two delta functions in the radiated field yield a sinusoidal response of the form

$$|e(\alpha)| = |1 - \exp(i\alpha)| = \sqrt{2(1 - \cos \alpha)} \qquad (3.21)$$

where $\alpha = kC$ and C is the circumference of the loop.

The convolution of the sinusoid component with a sine-wave excitation is more complicated but can be expressed in terms of elementary integrals as follows. Defining a single cycle sine wave function as

$$f(\theta) = \begin{cases} \sin \theta & 0 \leq \theta \leq 2\pi \\ 0 & \end{cases} \qquad (3.22)$$

we wish to find the convolution of this function with a harmonic function, i.e. we need to evaluate an integral of the form

$$g(\alpha, m) = \int_0^{2\pi} \sin \theta \, \sin(m\theta + \alpha) \, d\theta \qquad 0 \leq \alpha \leq 2\pi, \ 0 \leq m \leq \infty \qquad (3.23)$$

By using the trigonometric identities

$$2 \sin A \sin B = \cos(A - B) - \cos(A + B)$$

$$\cos(A + B) = \cos A \cos B - \sin A \sin B$$

$$\cos(A - B) = \cos A \cos B + \sin A \sin B$$

we can express this integral as the sum of four terms

$$g(\alpha,m) = \frac{\cos\alpha}{2}\left[\int_0^{2\pi}\cos(1-m)\theta\,d\theta - \int_0^{2\pi}\cos(1+m)\theta\,d\theta\right]$$

$$+\frac{\sin\alpha}{2}\left[\int_0^{2\pi}\sin(1-m)\theta\,d\theta + \int_0^{2\pi}\sin(1+m)\theta\,d\theta\right] \tag{3.24}$$

These can be easily integrated using

$$\int_0^{2\pi}\sin ax\,dx = \frac{1}{a} - \frac{\cos 2\pi a}{a}$$

$$\int_0^{2\pi}\cos ax\,dx = \frac{\sin 2\pi a}{a} \tag{3.25}$$

The result is that the sine-wave component yields the following convolution as a function of two parameters m and α, where m is the ratio of the drive angular frequency to β ($m > 0$) and α is the time-shift parameter of the convolution.

$$g(\alpha,m) = \frac{\cos\alpha}{2}\left\{\frac{\sin[2\pi(1-m)]}{1-m} - \frac{\sin[2\pi(1+m)]}{1+m}\right\}$$

$$+\frac{\sin\alpha}{2}\left\{\frac{2}{1-m^2} - \frac{\cos[2\pi(1-m)]}{1-m} - \frac{\cos[2\pi(1+m)]}{1+m}\right\} \tag{3.26}$$

If $m = 1$ this reduces to the special form

$$f(\alpha,m = 1) = \pi\cos\alpha \tag{3.27}$$

If we evaluate this function for $0 < m < 5$ and for each value of m let α vary such that $0 < \alpha < 2\pi$ and then plot the magnitude of the peak value of the resultant, we obtain the variation of radiated field with ω shown in Figure 3.21, which shows that the loop radiates efficiently only when the loop circumference is close to a wavelength. At higher frequencies the single-cycle sinusoid of the loop impulse response contains more and more cycles of the drive signal, which tend to average to zero.

- At low frequencies the radiation tends to zero (as it does for the dipole), since the convolution of the loop impulse response with a constant is zero. Note that this does not mean that the current on the loop is zero at

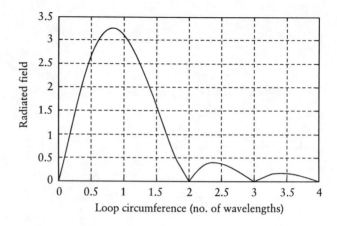

Figure 3.22 Feed-point current for a loop antenna as a function of frequency.

$\omega = 0$. Direct current can exist but it does not radiate. Indeed, whenever the current is uniform around the loop then, by symmetry, the field on the axis must be zero.

The input drive current for sinusoidal excitation is a function of frequency and has the form

$$i(t) = \sin(\omega t) + \sin(\omega t + \alpha) \tag{3.28}$$

where $\alpha = 2\pi rk$. The modulus of the associated complex phasor expression is given by

$$|i(\alpha)| = \sqrt{2(1 + \cos\alpha)} \tag{3.29}$$

which is shown in Fig. 3.22 for $0 < 2\pi r < 2\lambda$. We see that the current is maximum at zero frequency, i.e. the input impedance achieves a minimum.

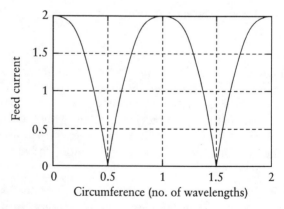

Figure 3.22 Feed-point current for a loop antenna as a function of frequency.

This is as expected, since at $\omega = 0$ the loop has zero resistance to current (in the absence of ohmic losses). Again, however, as we saw with the dipole, the situation changes as we increase the frequency.

The current achieves a minimum and the input impedance a maximum when the circumference is half a wavelength. The input impedance then falls to a minimum when the loop is one wavelength in circumference. For this reason the loop is most efficient for a full-wavelength resonance, which is very different behaviour to that observed for the dipole. We see that these differences can be traced to the polarity of the various contributions to the impulse response.

We can obtain an interesting variant on the circular loop by winding the wire into a helix and forcing the current to execute a helical path, thus forming a travelling wave in the z direction. Such antennas are very much more efficient than the loop and radiate circular polarization, as we now show.

3.5 Helical antennas

An interesting variant on the circular loop is to force the current to undergo a helical path in space by winding the wire into the form of a helix (either right or left handed), as shown in Figure 3.23. Such antennas are widely used for telemetry and space communications since they combine light weight, robust design and circular polarization. The latter is required for satellite communications, since propagation through the part of the Earth's atmosphere known as the "ionosphere" can rotate the plane of linear polarizations and so cause a mismatch at a linearly polarized receiving antenna. Circular polarization is invariant to such rotations (there is a change only in the phase of the wave).

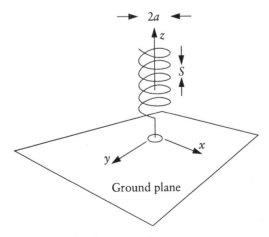

Figure 3.23 Geometry of a helical or spiral antenna.

The parametric equations for a source point on the helix are of the form

$$x = a\cos\beta t$$

$$y = \pm a\sin\beta t \tag{3.30}$$

$$z = \frac{S}{2\pi}\beta t$$

where S is the turn spacing (pitch of the helix), a is the helix radius, t is the time and β is the effective angular frequency. The plus sign indicates a right-handed helix and the minus a left-handed one.

The effective angular velocity is given by $\beta=2\pi/\tau$, where τ is the transit time around one turn of the helix, given by $\tau=L/c$. L is the length of wire in one turn of wire and is related to the pitch S and radius a by the triangle shown in Figure 3.24, from which we see that the effective angular velocity is given by $\beta=(c/a)\cos\gamma$. We obtain the limiting case of the loop as γ tends to zero, and a straight wire as γ tends to 90°.

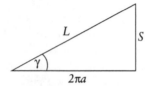

Figure 3.24 Triangle showing the geometry of one turn of the helix.

A full electromagnetic analysis of a helical structure such as this is quite complicated (see *Antennas* by Kraus, in suggestions for further reading given at the end of this chapter) and yields some very interesting results (for example, the phase velocity of a wave propagating along such a structure is very different to the free-space velocity of light, c). We can, however, gain some insight into the very important properties of the helix by attempting to use our acceleration principles and assuming that the current waves propagate at c.

There are three aspects to be considered in the design of such antennas.

Charge acceleration model

The charge acceleration is obtained by differentiating the parametric equations twice with respect to t. As for the loop antenna, this results in sinusoidal components for the acceleration vector which becomes

$$\mathbf{a} = -qa\beta^2(-\cos\beta t,\ \sin\beta t,\ 0) \tag{3.31}$$

Furthermore, we have seen from the case of the loop that such a radiation process is only efficient for sinusoidal excitation when the loop circumference approximately equals one wavelength. This is indeed the case for practical helical antennas. However, such antennas can operate over a relatively wide bandwidth, typically such that $0.75 < 2\pi a/\lambda < 1.333$.

Propagation of acceleration components to the observation point P

The propagation integral is slightly more difficult to evaluate than for the loop. To simplify matters we consider an observation point along the z axis with position vector $\mathbf{r} = (0, 0, r_p)$. It turns out that resonant helical antennas radiate most efficiently in this direction, i.e. along the axis of the helix.

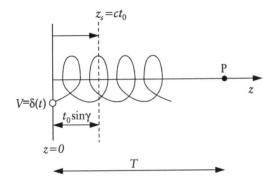

Figure 3.25 Helix propagation model along the z axis.

To obtain an approximate impulse response we need to consider the delta function excitation as shown in Figure 3.25. The impulse travels along the helix as a current wave with velocity c and we consider the total field at point P on the z axis as shown. This field $f(t)$ will be given by a sum of accelerations along the length of the helix. To determine the form of this sum or integral, consider the following argument.

The propagation time for a source at the origin 0 is T seconds. Now consider a delta function pulse moving along the helix with speed c. If the "helix time" is t_0 then the time of arrival t' at P is:

$$t' = t_0 + (T - t_0 \sin \gamma) \qquad (3.32)$$

where t_0 is the time to the launch point at $z = ct_0$ and $(T - t_0 \sin \gamma)$ is the time for propagation from z to P.

We wish to invert this to consider the source (t_0) of a fixed observation time t' at P. By simple rearrangement we obtain:

$$t_0 = \frac{t' - T}{(1 - \sin \gamma)} \qquad (3.33)$$

At this time the source has an x component of $\cos \beta t_0$ so the resultant field at P will have the form:

$$\cos\left(\frac{\beta t'}{1 - \sin \gamma}\right) \tag{3.34}$$

If we just consider this cosine component of the acceleration and integrate along the length of the helix, we obtain an expression of the form:

$$f_p(t) = \int_0^{\frac{NL}{c}} \delta(t - t_0 + t_z) \cos \beta t_0 \, dt_0 = \cos\left(\frac{\beta t}{1 - \sin \gamma}\right)\Bigg|_{t=0}^{t=\frac{2\pi Na}{c \cos \gamma}} \tag{3.35}$$

where N is the number of turns of the helix, and from the geometry of the helix

$$\frac{NL}{c} = \frac{2\pi Na}{c \cos \gamma} \tag{3.36}$$

$t_0 = s/c$, and s is the distance along the helix. The time t_z is given by $t_z = t_0 \sin \alpha$ and is the time taken for light to propagate the distance z_s.

A similar expression is obtained for the sinusoidal component of acceleration. The final radiated impulse response of a right-handed N-turn helix on the axis then has the general form

$$h(t) = -q\beta^2 a\left[\cos\left(\frac{\beta t}{1 - \sin \gamma}\right)\Bigg|_{t=0}^{t=\frac{2\pi Na}{c \cos \gamma}} \mathbf{i} + \sin\left(\frac{\beta t}{1 - \sin \gamma}\right)\Bigg|_{t=0}^{t=\frac{2\pi Na}{c \cos \gamma}} \mathbf{j}\right] \tag{3.37}$$

which for large N represents a circularly polarized wave.

Figure 3.26 shows the qualitative radiation behaviour of a three-turn helix with radius $a = 1\,\mathrm{m}$ and pitch angle $\gamma = 18°$. The upper plot is the variation of the phase difference between the x and y components of the radiated field. The lower plot is of the maximum radiated field strength. Both are plotted as a function of the frequency of the excitation relative to a reference frequency of

$$\omega_{ref} = \frac{c \cos \gamma}{a(1 - \sin \gamma)} \tag{3.38}$$

We see that only if the relative drive frequency is close to unity do we obtain a large radiated field and that over this range the antenna radiates circular polarization (i.e. a 90° phase shift between x and y with equal amplitudes).

The plots were obtained using our approximation for $h(t)$ as derived in Eq. (3.37) and numerical convolution with a sinusoidal excitation of the appropriate frequency. The plots were then obtained by measuring the peak value of the resultant convolution with the x and y components of $h(t)$.

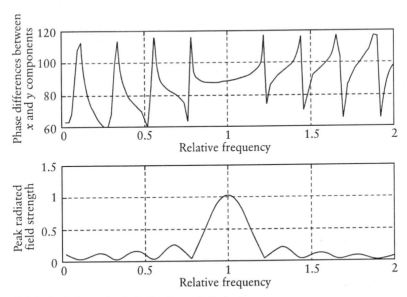

Figure 3.26 Radiation and polarization from a helical antenna.

A more exact analysis shows that the axial ratio for an n-turn helix operating at $\omega = \omega_{ref}$ is given by

$$AR = \frac{2n+1}{2n} \tag{3.39}$$

which for our three-turn example gives $AR = 1.17..$

Again, as in the case for the loop and dipole, our simple theory provides the correct qualitative behaviour and permits us to see more clearly how the resonant behaviour of a helix is related to the structure of its impulse response.

Impedance properties of helical antenna

The helix antenna has some rather unusual impedance properties that we, nevertheless, can explain using our simple model of charge acceleration. Note that the open-circuit termination at the end of the helix (Fig. 3.23) has only a small effect since, for a resonant helix with several turns, most of the energy is radiated by the curved current path of the helix. Thus we have the rather remarkable result that although the helix has no matched termination

81

(i.e. we expect it to support standing waves), it behaves very like a travelling wave antenna.

Another very important consequence of generating radiation from the whole antenna structure is that the design is not critically dependent on the antenna dimensions. This makes the design and manufacture of helical antennas relatively robust and straightforward.

In summary, we have seen that we can explain many of the important general features of antenna behaviour using simple impulse response approximations. For more qualitative information about antennas, we need to make use of detailed mathematical analysis based on rigorous solutions to ME or to make use of numerical methods suitable for implementation on a digital computer. In the following chapter we consider the latter set of methods and see how they can be used to extract important design information about antenna performance. We shall see that in order to do this we must first return to our basic ideas of wave propagation using the advection equation.

Suggestions for further reading

Most books treat the subject of wire antennas from the point of view of frequency-domain concepts and so emphasize different aspects of the calculation of radiation and impedance properties than those discussed in this chapter. The following books provide interesting developments of the fundamental properties of such antennas. Many other texts are available containing design information on specific wire antenna structures. However, very often these do not provide a clear physical explanation of the antennas' operation.

Collin, R. E. 1985. *Antennas and radiowave propagation.* New York: McGraw Hill.

Harrington, R. F. 1961. *Time harmonic electromagnetic fields.* New York: McGraw-Hill.

Kraus, J. D. 1988. *Antennas.* New York: McGraw Hill.

Landt, J. A. & E. K. Miller 1974. Short pulse response of a circular loop. *IEEE Transactions* AP-22, 114–16.

Lee, K. 1984. *Principles of antenna theory.* New York: John Wiley.

Mott, H. 1992. *Antennas for radar and communications: a polarimetric approach.* New York: John Wiley.

Papas, C. H. 1988. *Theory of electromagnetic wave propagation.* New York: Dover Press.

Rumsey, V. H. 1966. *Frequency independent antennas.* New York: Academic Press.

Problems

3.1 Obtain an expression for the phasor representing radiation from a linear dipole of length L at 30° from broadside. Hence find the lowest frequency f for which the radiation achieves a maximum in this direction. Compare this with the frequency for which the broadside radiation is maximum.

3.2 By considering current wave propagation, prove that the radiated impulse response for a horizontal travelling wave antenna is of the form shown in Figure 3.13. Hence calculate the radiated impulse response in a direction $\theta = 60°$ if the antenna has a length of 10 m. At what frequency is this radiation maximum?

3.3 A centre-fed wire antenna of length $L = 2$ m is fed with a step pulse of rise time 500 ps. Sketch the form of the radiated field at the broadside (assume that the wire is impedance matched at the feed point).

3.4 Find an expression for the input impedance and broadside radiated field as a function of frequency for the V-antenna shown below. The broadside direction bisects the angle AOB and $L_2 < L_1$. The feed is at point O and you may assume it is matched to the antenna.

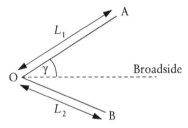

3.5 Consider the problem of finding the radiation from the corner of a rectangular loop as shown below.

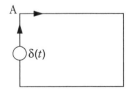

A canonical problem with such an antenna is to calculate the radiation by charge moving through 90°, i.e. around a corner such as A. We can model the position vector of a point moving around such a corner as

$$\mathbf{r} = \frac{1}{2}\big[[1 + \tanh(at)]\,\mathbf{i} + [1 - \tanh(at)]\,\mathbf{j}\big]$$

$$\tanh(x) = \frac{\exp(x) - \exp(-x)}{\exp(x) + \exp(-x)}$$

Sketch the form of the x and y components of the vector $\mathbf{r}$ and hence determine the influence of the parameter a. Using this model, calculate the acceleration vector and hence determine the radiation from such a corner in the direction $\mathbf{r}_p = (0, 0, 1)$.

3.6 A helical antenna is to be designed for use at a resonant frequency $f = 500\,\text{MHz}$. The axial ratio of the polarization ellipse must be better than 1.02. If the antenna radius is to be restricted to 0.5 m, calculate the helix angle, minimum number of turns and overall length of the antenna. What total length of wire will be required to manufacture the antenna?

Chapter 4

Computational methods for wave propagation

4.1 Finite difference approximations

In Chapter 1 we used our intuitive understanding of the local nature of wave motion to generate mathematical equations governing all types of wave. In this chapter we reverse this process – we begin with the differential equations and derive approximations in the form of difference equations. The great advantage of using these difference equations is that they can be solved on a digital computer to provide *approximate* solutions to the differential equations. In this way we can employ computers to solve wave propagation, generation and scattering problems.

We saw in Chapter 1 that an important example of a wave function is the harmonic or sinusoidal wave, represented in one dimension by the function $f(z,t) = \sin(vt - z)$, where v is the wave velocity. If we form partial derivatives with respect to space and time we obtain the advection equation as

$$\left.\begin{aligned}\frac{\partial f}{\partial t} &= v\cos(vt - z) \\ \frac{\partial f}{\partial z} &= -\cos(vt - z)\end{aligned}\right\} \quad f_t + vf_z = 0 \tag{4.1}$$

and, as observed in Chapter 1, any mathematical function which satisfies this equation is also a wave propagating in the $+z$ direction.

One of the simplest numerical approaches to solving the advection equation is to use a finite difference approximation to the partial derivatives and obtain an explicit time-stepping solution.

The finite difference technique begins with a discretization of the solution space (in this case the zt plane as shown in Fig. 4.1) where consideration is given only to space-time nodes located by a pair of indices j and n. We represent the space time value of a function at each node by the subscript/superscript notation f_j^n.

The basic idea is to approximate the derivative of f at a point, formally defined as

85

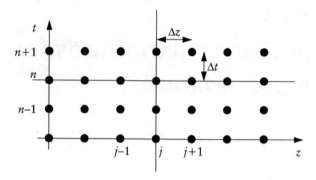

Figure 4.1 Discretization of space and time for the finite difference method.

$$\frac{df}{dz} = \lim_{h \to 0} \frac{f(a+h) - f(a)}{h} \tag{4.2}$$

by one of the following finite difference approximations

$$\frac{df}{dz} = \begin{cases} \dfrac{f(a+\Delta) - f(a)}{\Delta} & \text{forward difference} \\[2mm] \dfrac{f(a) - f(a-\Delta)}{\Delta} & \text{backward difference} \\[2mm] \dfrac{f(a+\Delta) - f(a-\Delta)}{2\Delta} & \text{central difference} \end{cases} \tag{4.3}$$

where Δ is some small but finite increment (such as Δz or Δt in Fig. 4.1). We can extend this idea in a straightforward manner to approximate second- and higher-order derivatives. For example, the second-order derivative can be approximated by the finite difference of first-order forward and backward differences to obtain

$$\begin{aligned} \frac{d^2 f}{dz^2} &= \frac{1}{\Delta} \cdot \frac{f(a+\Delta) - f(a) - f(a) + f(a-\Delta)}{\Delta} \\[2mm] &= \frac{f(a+\Delta) - 2f(a) + f(a-\Delta)}{\Delta^2} \end{aligned} \tag{4.4}$$

To see how this can be used to obtain a numerical solution to the advection equation of Eq. (4.1), consider replacing the time partial derivative by a forward difference and the space derivative by a central difference:

$$\frac{\partial f}{\partial t} \approx \frac{f_j^{n+1} - f_j^n}{\Delta t}$$

$$\frac{\partial f}{\partial z} \approx \frac{f_{j+1}^n - f_{j-1}^n}{2\Delta z} \tag{4.5}$$

By substituting these into the advection equation and rearranging, we can express the up-dated function value f_j^{n+1} in terms of values at earlier space time points:

$$f_j^{n+1} = f_j^n - \frac{v\Delta t}{2\Delta z}\left(f_{j+1}^n - f_{j-1}^n\right) \tag{4.6}$$

This is a very important relation. It permits us to time step through a solution for the function f by starting at time $t = 0$ and applying an explicit formula for all j at the first time step $n = 1$. Then we repeat the exercise for $n = 2$, etc. We can represent the local nature of this updating process by considering evaluation of the function at node n,j as shown in Figure 4.2.

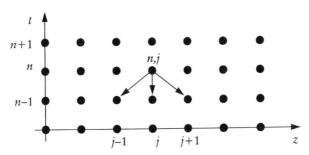

Figure 4.2 Local dependence in explicit finite difference techniques.

We would begin such a scheme by employing some initial excitation function. This excitation can be any function of time $g(t)$ and our updating formula will show how such a disturbance propagates as a wave. In the example code shown in Appendix B, Section 1, the excitation function is a Gaussian pulse. The code is written in MATLAB and the reader is encouraged to run the code and view the output

Unfortunately, whatever choice is made for Δt and Δz in Eq. (4.6), the function f grows exponentially and fails to model the propagation of a Gaussian pulse. In mathematical terms the method is unconditionally unstable and all seems lost. It seems from this example that it is impossible to model wave propagation using finite difference approximations.

We shall see that, while caution is required, we can modify our simple model to obtain a stable finite difference model. The lesson should not be lost however: straightforward discretization of differential equations can lead to models which are next to useless.

We can investigate the root of this instability by considering the special case when the wave function is a Fourier (or harmonic) mode defined as:

$$f_j^n = \exp(i\theta) = \exp\left[i(\omega n\Delta t - kj\Delta z)\right] = \cos\theta + i\sin\theta \qquad (4.7)$$

where $i^2 = -1$. The imaginary part of this expression is simply our harmonic wave function. When we substitute this in the difference equation, we obtain

$$f^{n+1}\exp(-ikj\Delta z) = f^n\exp(-ikj\Delta z) -$$
$$f^n\frac{v\Delta t}{2\Delta z}\left[\exp(-ik(j+1)\Delta z) - \exp(-ik(j-1)\Delta z)\right] \qquad (4.8)$$

which we can simplify by factoring out the term $\exp(-ijk\Delta z)$ and using the fact that $\exp(i\theta) - \exp(-i\theta) = 2i\sin\theta$ to obtain

$$f^{n+1} = f^n\left[1 + i\frac{v\Delta t}{\Delta z}\sin(k\Delta z)\right] \qquad (4.9)$$

This simple difference equation is of the form $f^{n+1} = gf^n$, where g is a constant. This equation has very different properties if g has a modulus value greater or less than unity. If $|g| > 1$ then we see that repeated application of the update equation will lead to magnification and instability. Only if $|g| < 1$ will this equation remain stable.

In our example, g is a complex number given by

$$g = 1 + i\alpha \quad \text{where} \quad \alpha = \frac{v\Delta t}{\Delta z}\sin(k\Delta z) \qquad (4.10)$$

The modulus of this number is

$$|g|^2 = gg^* = 1 + \alpha^2 \geq 1 \qquad (4.11)$$

which is always greater than 1 and we see that, regardless of how small we make the space and time steps, our simple finite difference model will always be unstable. This is surprising: intuition tells us that, as we make the elements Δt and Δz smaller, the solution of the difference equation should approach the solution of the differential equation. It is a fact of life that this is not the case.

The cure to this instability problem is remarkably simple but introduces a cost. The cost is that, instead of our advection equation, we end up solving a modified equation with an extra pair of second derivatives. This equation is stable but shows dispersion, i.e. each Fourier mode governed by a choice of k has a corresponding value of ω which depends on k, i.e. the velocity of the wave $v = \omega/k$ depends on k. This can be a serious problem since a pulse excitation contains many k components (given by Fourier analysis), and as they propagate on the grid they travel with different velocities.

The resultant spreading of the pulse as it propagates is called *dispersion*. We shall see that there are some simple rules for minimizing this dispersion. Note that this is not a physical dispersive effect (as occurs, for example, when light is split into different colours by a prism). For this reason we specify that there is *numerical* dispersion on a finite difference grid.

We can generate the stabilization method by considering the form of g, the amplification factor given by $g = (1+i\alpha)$. We want to make changes so that g has a modulus less than unity for some values of α. The problem lies with the unity term on the right-hand side of Eq. (4.11); if we could replace that term with a trigonometric factor such as a cosine factor, then there would be a possibility that the modulus could be less than unity since

$$g = \cos\theta + i\beta\sin\theta$$
$$gg^* = \cos^2\theta + \beta^2\sin^2\theta$$

(4.12)

which is less than one if $\beta < 1$. The only remaining problem is how to translate the requirement for the cosine into a modification of the difference equation. We can do this by noting that $\cos\theta = \frac{1}{2}(\exp(i\theta) + \exp(-i\theta))$. Translating this into a general condition for our difference equation we see that we must take a spatial average of the right hand side to obtain a new equation

$$f_j^{n+1} = \frac{1}{2}\left(f_{j+1}^n + f_{j-1}^n\right) - \frac{v\Delta t}{2\Delta}\left(f_{j+1}^n - f_{j-1}^n\right)$$

(4.13)

This equation is stable as long as $\beta = (v\Delta t)/(\Delta z) \le 1$. Appendix B, Section 2 gives a MATLAB code which implements this modified advection equation. Such averaging methods are called "Lax methods" in numerical analysis.

If this code is run for 30 space segments and 100 time steps, we obtain the space time plot (the array z) shown in Figure 4.3 where we see the Gaussian pulse propagating smoothly across the discrete mesh. As long as $\beta < 1$ (which amounts to making the correct choice for the space and time steps) we have a stable model.

We can rewrite our Lax difference scheme in the form

$$\frac{1}{2}\left(f_j^{n+1} - f_j^{n-1}\right) + \frac{1}{2}\left(f_j^{n+1} - 2f_j^n + f_j^{n-1}\right) =$$
$$\frac{1}{2}\left(f_{j+1}^n - 2f_j^n + f_{j-1}^n\right) - \frac{v\Delta t}{2\Delta z}\left(f_{j+1}^n - f_{j-1}^n\right)$$

(4.14)

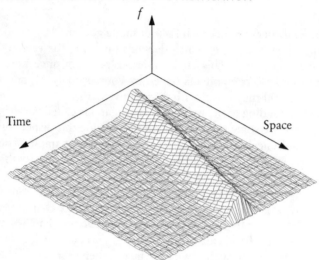

Figure 4.3   Space/time mesh for Gaussian pulse propagation.

from which we see that by using the Lax scheme we are actually solving a differential equation of the form

$$\frac{\partial f}{\partial t} + \frac{\Delta t}{2}\frac{\partial^2 f}{\partial t^2} - \frac{\Delta z^2}{2\Delta t}\frac{\partial^2 f}{\partial z^2} + v\frac{\partial f}{\partial z} = 0 \tag{4.15}$$

where the first and last terms are identified as elements of our advection equation. The second-order space derivative is the term which provides stability for our method. The price we pay for this stability is the presence of dispersion, as mentioned above.

This dispersion is more of a problem for large wave numbers (small wavelength) and so we must take care when exciting a finite difference grid to model wave propagation that the pulse width is greater than about 10 time steps so that it does not contain Fourier modes with high wave numbers. We must also avoid sudden time discontinuities such as steps and impulses. These contain high frequencies and have high wave number components which will undergo numerical dispersion on the grid.

This need to use space and time steps much smaller than the pulse width is a major practical limitation in finite difference techniques. The prediction of high frequency effects requires short pulses and even shorter time steps. In order to make predictions over several transit times of the object we consequently need to store **E** and **H** values over a very dense grid, which uses up a great deal of computer memory; in fact, computer memory limitations restrict the high frequency application of such finite difference techniques.

So far we have considered only the possibility of waves propagating in one direction. We now consider how to extend the above ideas to the full wave equation.

4.2 Finite difference for the wave equation

As discussed in Chapter 1, the wave equation (for one spatial dimension) is formed as the product of two advection equations for waves travelling in plus or minus directions. The resulting second-order equation is of the form

$$\frac{\partial^2 f}{\partial z^2} - \frac{1}{v}\frac{\partial^2 f}{\partial t^2} = 0 \tag{4.16}$$

There are two methods for turning this equation into a finite difference model. The first is to directly replace the second-order derivatives by second-order differences and rearrange terms to obtain the following explicit time-stepping algorithm

$$f_j^{n+1} = \gamma^2\left(f_{j+1}^n + f_{j-1}^n\right) + 2\left(1 - \gamma^2\right)f_j^n - f_j^{n-1} \tag{4.17}$$

where $\gamma = v\Delta t/\Delta z$, and the space/time grid for the function f has the form shown in Figure 4.4. This method is stable as long as $\gamma < 1$.

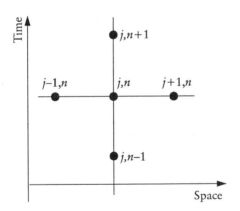

Figure 4.4 Indexing for the finite difference form of the wave equation.

The second method (and the one favoured for our purposes) is called a "leap-frog algorithm". In this case we replace our single second-order Eq. (4.16) for f by a pair of first-order advection equations. The price we pay for this is the need to introduce a second function g, such that the coupled pair

$$\frac{\partial f}{\partial t} = v\frac{\partial g}{\partial z}$$

$$\frac{\partial g}{\partial t} = v\frac{\partial f}{\partial z} \tag{4.18}$$

is equivalent to our single second-order wave equation (if we differentiate the first equation with respect to t and the second with respect to z we obtain our second-order wave equation).

To discretize this pair of equations we apply our finite difference approximations, calculating the two functions f and g at *alternate* space/time points so we obtain the interleaved space/time nodes shown in Figure 4.5.

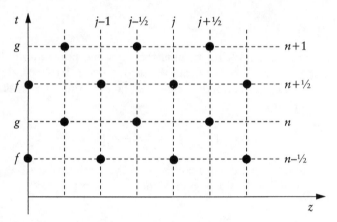

Figure 4.5 Space and time indexing in the leap-frog method.

This method is preferred in electromagnetics since our two functions f and g can be associated directly with the electric and magnetic fields **E** and **H**. As we saw in Chapter 1, both **E** and **H** are involved in the electromagnetic wave process. The leap-frog scheme then makes explicit use of the coupling between **E** and **H**.

The first person to detail the finite difference method for Maxwell's equations was Yee in 1966, hence the discretization of these equations is called the "Yee algorithm". It is the basis for most time domain finite difference methods used today.

To illustrate how the leap-frog algorithm may be applied to electromagnetic problems, consider the case of using Maxwell's equations for a description of wave propagation in one dimension (the z direction). In this case the two Maxwell curl equations reduce to equations of the form

$$\frac{\partial H_y}{\partial t} = -\frac{1}{\mu}\frac{\partial E_x}{\partial z}$$

$$\frac{\partial E_x}{\partial t} = -\frac{1}{\varepsilon}\left(\frac{\partial H_y}{\partial z} + \sigma E_x\right) \qquad (4.19)$$

If we now use the first-order difference approximations,

$$\frac{\partial \mathbf{H}}{\partial t} \approx \frac{H_j^{n+\frac{1}{2}} - H_j^{n-\frac{1}{2}}}{\Delta t} \qquad \frac{\partial \mathbf{E}}{\partial z} \approx \frac{E_{j+\frac{1}{2}}^n - E_{j-\frac{1}{2}}^n}{\Delta z}$$

$$\frac{\partial \mathbf{E}}{\partial t} \approx \frac{E_{j-\frac{1}{2}}^{n+1} - E_{j-\frac{1}{2}}^n}{\Delta t} \qquad \frac{\partial \mathbf{H}}{\partial z} \approx \frac{H_j^{n+\frac{1}{2}} - H_{j-1}^{n+\frac{1}{2}}}{\Delta z}$$

(4.20)

and rearrange the resulting equations, we obtain the following explicit time-stepping procedures for the **E** and **H** fields

$$H_j^{n+\frac{1}{2}} = H_j^{n-\frac{1}{2}} - \frac{\Delta t}{\mu \Delta z}\left(E_{j+\frac{1}{2}}^n - E_{j-\frac{1}{2}}^n \right)$$

$$E_{j-\frac{1}{2}}^{n+1} = \left[\frac{1 - (\sigma \Delta t / 2\varepsilon)}{1 + (\sigma \Delta t / 2\varepsilon)} \right] E_{j-\frac{1}{2}}^n + \frac{\Delta t}{\varepsilon \Delta z \left[1 + (\sigma \Delta t / 2\varepsilon) \right]}\left(H_j^{n+\frac{1}{2}} - H_{j-1}^{n+\frac{1}{2}} \right)$$

(4.21)

Note that the fields are evaluated at alternate time steps, as required in the leap-frog approach. When coding this difference scheme on a computer we must evaluate all **H** field nodes in the z direction at time step t then evaluate all **E** field nodes at time step $t+1$, all **H** field nodes again at $t+2$, etc. A MATLAB implementation of this differencing scheme is shown in Appendix B, Section 3. Note how the space and time loops are staggered, as required in the differencing scheme.

If we run the above code with parameters $ns = 41$ and $nt = 100$, we obtain the **H** field space/time plot shown in Figure 4.6. Note that in this example we

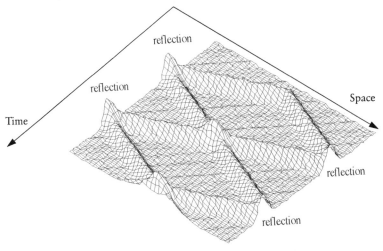

Figure 4.6 Space/time mesh for the wave equation with zero field boundary conditions.

feed the array from the centre to generate two waves propagating in the $+z$ and $-z$ directions. The waves propagate away from the source point until they reach the edges of the array. We see that the waves are then reflected back into the computational domain. Unfortunately these are not physical reflections but artefacts, due to the fact that our array must be bounded inside the computer. This undesirable situation can be removed by using absorbing boundary conditions at the edges of the array. We now turn to consider such conditions.

4.3 Absorbing boundaries

The above differencing scheme is appropriate for all interior space–time mesh points but, as pointed out, we require a special procedure for points at the ends of the array, where calculation of one field component requires a knowledge of the coefficients of the other which lie outside the array.

The simplest procedure is to set these unknown components to zero and use the same time-stepping formula for all points. However, this causes a serious problem: by enforcing some components to zero we generate a perfect magnetic or electric conductor (depending on whether the boundary points are in the E or the H field (see Chapter 5)). Such perfect conductors act as perfect reflectors and the waves are reflected back into the computational space. As this reflection is not physically desirable, it generates an error in the computation of wave propagation and scattering. This problem is illustrated in the space-time plot shown in Figure 4.6, where all field components outside the array were set to zero.

This problem is a serious limitation of finite difference methods for modelling wave propagation. There has been a great deal of work focused on trying to solve this problem. The basic idea behind such schemes is to "guess" the values of the external unknown points so as to "fool" the algorithm into thinking, at least locally, that the boundary does not exist. If we can achieve such a condition then waves incident on the edges are absorbed with zero reflection. Such methods are known as "absorbing boundary conditions".

There are several sophisticated approaches to this problem but here we outline a simple method for the one-dimensional case which illustrates the principle of the method. We choose to absorb the H field but a similar argument can be used for the E field components.

From inspection of Eq. (4.21), our problem lies in the evaluation of the H field from the spatial derivative of the E field through Maxwell's equation. The values of the E field required to determine H lie outside the array and hence are unknown. We can solve this by replacing the Maxwell equation relating E and H by a new differential equation governing the local behaviour of H only.

Since we expect H to propagate as a wave it would seem appropriate to use an advection equation (a one-way equation will suffice since we know that the wave will be incident from the centre of the array). We then suggest that, locally at least, **H** must satisfy the equation

$$\frac{\partial h}{\partial t} = v \frac{\partial h}{\partial z} \qquad (4.22)$$

If we use first-order differences for this equation we obtain the following approximation

$$h_j^{n+\frac{1}{2}} = h_j^{n-\frac{1}{2}} + \frac{v \Delta t}{\Delta z} \left(h_{j-1}^{n-\frac{1}{2}} - h_j^{n-\frac{1}{2}} \right) \qquad (4.23)$$

If we also use the condition $v \Delta t = \Delta z$, then we see that this equation has a particularly simple form. The value of **H** on the boundary at the update time step is equal to the value from the previous time step at the neighbouring spatial node. In other words, we assume the **H** field propagates as a wave, moving one space step in one time step.

If we employ this equation on the boundaries and the full Maxwell equations at other points, we can achieve an absorbing boundary as shown in Figure 4.7. This space/time grid was obtained by running the MATLAB code for the leap-frog algorithm with $ns = 41$ and $nt = 100$.

Note that the pulse propagates away from the source at the centre of the array and propagates through the boundaries with zero reflection.

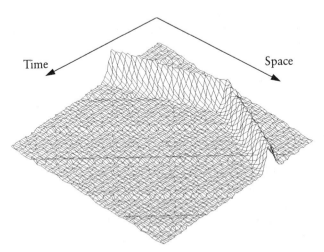

Figure 4.7 Space/time mesh for the wave equation with absorbing boundaries.

We now have a complete and stable numerical model for waves propagating in both the plus and the minus directions with velocity v, and have removed the problem of having only a finite computational zone inside the computer. This code can now be used to model wave problems where the wave velocity is a function of z. In such cases we shall see the familiar wave phenomena of reflection and interference.

4.4 Reflection and refraction of waves

Consider the simple but important problem of wave reflection from a layer of material with relative permittivity ε_r, conductivity σ and thickness d embedded in free space (represented as a computational zone of length L where $L \gg d$ as shown in Fig. 4.8).

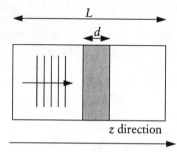

Figure 4.8 Geometry of wave reflection from slab of material of thickness d.

Absorbing boundaries must be placed at the left- and right-hand edges of the computational space L. Physically, if we have a wave incident on the slab from the left we expect to see reflected waves from the front and rear surfaces. These reflections will undergo two important processes inside the slab d. The first is multiple reflection as the wave propagates from side to side. The second is absorption of energy in the material. This absorption will only occur if the material conductivity σ is non-zero.

Appendix B, Section 4 gives a MATLAB code which uses the finite difference method to model the above problem of reflection from a slab in free space. The slab width is 40 space steps. The space-step size depends on the number of segments used to fill the 1 m computational space. The default number of segments is $ns = 51$. These numbers can be changed depending on the amount of computer memory available and the required speed of computation.

Figure 4.9 shows the space/time plot obtained for reflection of a pulse from a layer with permittivity $\varepsilon_r = 16$ and zero conductivity. The Gaussian pulse is launched from the top left-hand corner. It propagates and is reflected from the front surface of the layer. The reflected pulse propagates out of the computational space and is absorbed by the absorbing boundary conditions. The refracted pulse enters the slab and undergoes multiple reflections from the

96

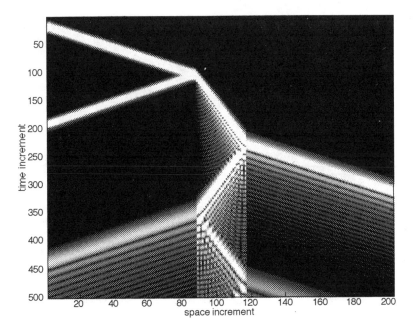

Figure 4.9 Space/time plot showing wave reflection and refraction from a dielectric layer with $\varepsilon_r = 16$, and $\sigma = 0$.

back and front surfaces of the slab. Note how the pulse disperses as it propagates. This is due to numerical dispersion on the grid and is not a real physical effect (see Eq. (4.15)). To reduce this dispersion we must employ a wider pulse width to reduce the high frequency pulse content.

Such internal multiple reflections as observed in Figure 4.9 can be used to design optical and microwave filters which have the useful property of passing only certain harmonic wave components with wave numbers which fall in a pass band of values determined by choosing ε_r and d according to wavelength. This pass band arises from the constructive interference of the various reflected waves inside the slab.

Although we have used the simple problem of a slab of constant material properties, it should be clear that, due to the local nature of the finite difference approximation, we can set the material constants to have any variation in space and time. This ability to model general problems is a great strength of the finite difference approach.

To illustrate what happens in a lossy medium we show in Figure 4.10 the space/time history for a pulse incident on the slab with permittivity $\varepsilon_r = 16$ and conductivity $\sigma = 0.25 \, \text{S m}^{-1}$. Note how this time the first pulse reflection is large and the component which propagates inside the slab is absorbed (the energy in the pulse being dissipated as heat or ohmic losses). As a result, the multiple reflections seen in Figure 4.9 are not so important.

97

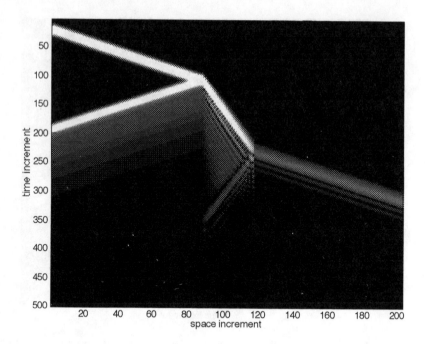

Figure 4.10 Space/time plot for wave interaction with absorbing media (ε_r = 16 and σ = 0.25).

This effect is used in the design of metal shields for electronic components. By using thin sheets of high conductivity material, any waves incident on the sheet from outside are reflected and absorbed before they can reach the internal electronics and cause interference by generating voltages and currents.

4.5 Integral equations and the method of moments

An alternative to employing finite difference methods for the modelling of wave effects is to use the concept of integral equations. In this case we express the solution to the wave equation $\Psi(r,t)$

$$\left(\nabla^2 - \frac{1}{v^2}\frac{\partial^2}{\partial t^2}\right)\Psi(r,t) = f(r,t) \qquad (4.24)$$

as a convolution integral of the source term $f(r,t)$ with the impulse response of the wave equation. In the context of wave equations, the impulse response is called a "Green's function" after the nineteenth century English mathematician George Green (1793–1841), whose studies of electricity and magnetism led him to many concepts and mathematical theorems which are now used in a wide range of physical disciplines.

To generate the Green's function for the wave equation, we must consider an impulse excitation in space and time so that our delta function excitation now corresponds to $\delta(r-r')\delta(t-t')$, where r' indicates the position and t' the time at which the impulse is applied. The response to such an impulse will be represented as $G(r-r';t-t')$. By definition, this function satisfies the following differential equation

$$\left(\nabla^2 - \frac{1}{v^2}\frac{\partial^2}{\partial t^2}\right)G(r-r';t-t') = \delta(r-r')\delta(t-t') \tag{4.25}$$

The function $G(r-r';t-t')$ is the Green's function for the time dependent wave equation. It gives the response at position r and time t due to a point source located at position r' and time t'. The various possible Green's functions arise by imposing boundary conditions on the wave components. We shall consider the simplest case where the wave is unbounded.

The main point to note is that if we can obtain this special function, we can then obtain the response for an arbitrary source $f(r,t)$ by a convolution integral (just as we did for wire antennas in Chapter 2), i.e. we obtain a general solution of Eq. (4.24) in the form of an integral as

$$\Psi(r,t) = \int\int\int\int f(r',t)G(r-r';t-t')\,dv\,dt' \tag{4.26}$$

where dv is a volume element in space. Notice that three of the integrations are concerned with spatial coordinates and one with time.

The formal derivation of the Green's function is a complex process, but we can arrive at a form for the function in unbounded media by simple intuition, as follows. The result is the same as that obtained by the more formal approach and introduces us to the important concept of retarded integrals.

We saw when considering the radiation from a point charge in Chapter 2 that radiated fields decay with distance as r^{-1}, where r is the radial distance from the point source. Hence it is reasonable for us to postulate the following form of the Green's function:

$$G(r-r';t-t') = \frac{1}{4\pi|r-r'|}G(t-t') \tag{4.27}$$

where the $1/4\pi$ scale factor is included to conform with Coulomb's law. Such a $1/r$ dependence is typical of all wave motions in three-dimensional space.

We can also obtain the form of the time-dependent term $G(t-t')$ by using the principle of causality. This states that the effect (the response) cannot precede the cause (the source). We used this principle when generating the fields from an accelerating charge. Here we use it to force the concept of a maximum velocity of propagation v and hence the idea that the point source cannot influence other points until after a time related to the distance separating the two points divided by the velocity of propagation. If we assume the delta function excitation propagates without distortion, the final form of our postulated Green's function must then be

$$G(r - r'; t - t') = \frac{1}{4\pi|r - r'|} \delta\left(t - t' - \frac{|r - r'|}{v}\right) \qquad (4.28)$$

This is the same form of the Green's function as obtained by more formal mathematical procedures which involve Fourier transforms and some advanced calculus. We have developed it based on simple physical principles which are valid for all time-dependent wave processes.

If we substitute this form for the Green's function into our convolution integral and use the properties of the delta function, we obtain an integral relation between the source and wave function as

$$\Psi(r,t) = \frac{1}{4\pi} \iiint \frac{f\left(r', t - |r - r'|/v\right)}{|r - r'|} \, dv \qquad (4.29)$$

This is called the "retarded solution" to the wave equation. It is an integration over the sources not at time t but at a time $|r - r'|/v$ seconds earlier. This formal solution is the basis for a wide range of integral equation solution methods.

In a direct sense this integral can be used to find the field Ψ radiated by sources f (such as accelerating charges) by employing a summation over these sources at the appropriate times. However, more often use is made of this integral relation in another way, i.e. to solve for the sources f given the field values Ψ. In this case the unknown function is *inside* the integral and to find a solution we must solve a so-called "integral equation".

Integral equations are widely used for the solution of antenna and wave-scattering problems. Their primary advantage over the finite difference methods is their more efficient use of computer memory, potential for improved accuracy and the fact that they do not require explicit implementation of absorbing boundary conditions. Their prime disadvantages are their relative mathematical complexity and restriction to a class of problems for which the Green's functions are known.

The integral equations are formulated by employing three known facts:
- The field values $\Psi(r,t)$ are usually specified at some special points (say at the feed point of the antenna).
- The retarded solution must apply at *all* points in space and time.
- That on some known surface in space the fields must satisfy a prescribed

boundary condition (for example, that the tangential electric fields must be zero at the surface of a perfect conductor).

In electromagnetics, the boundary conditions can be applied to the electric or magnetic fields and hence there are two classes of integral equation, depending on which boundary condition we apply: the magnetic field integral equation (usually abbreviated to MFIE) and the electric field integral equation (EFIE).

Surprisingly, these three conditions are sufficient for us to obtain a numerical solution to the problem of finding the source terms everywhere in space and time. This is achieved using a technique known as the "marching-in-time algorithm".

To illustrate the time marching solution method we do not consider the EFIE or MFIE in any detail but instead employ a generic form of a time domain integral equation which can be written in the form

$$\Psi(r,t) = \Psi_i(r,t) + \int_V dr' \int_0^{t-\frac{R}{v}} dt' \, G(r-r';t-t') \Psi(r',t') \tag{4.30}$$

where $\Psi_i(r,t)$ is the (known) incident field and $\Psi(r,t)$ is the unknown function to be determined. Note that the unknown $\Psi(r,t)$ appears inside the integral as well as on the left-hand side of the equation.

The secret of the time-marching method is to employ the retarded (or delayed) time solution inside the integral on the right-hand side and to march forward in time, calulating the present state of the function $\Psi(r,t)$ using only past values, which have been calculated at earlier time steps.

The basic numerical approach is to discretize space and time, approximate the integral by a finite sum and invoke the equality sign of the above equation at appropriate space–time points. This process of solving an integral equation by approximating the integral by finite sums is called the *method of moments*. Since it results in a discrete set of linear equations which can be solved on a computer, this method is very important in the numerical solution of electromagnetic and other wave problems. The only drawback is that as the problem becomes large so the number of equations becomes very large and the computational time and memory storage becomes prohibitive.

To see how such a scheme works, we begin by constructing a uniform spatial grid $\{r_\alpha\}$ with mesh size h and take time in discrete steps $t = n\Delta t$ where $n = 0, 1, 2, 3, \ldots$. In order to avoid the requirement to solve a matrix equation for the unknowns, we choose the time step Δt such that

$$\Delta t = \frac{\min(R_{\alpha\alpha'})}{v} \qquad R_{\alpha\alpha'} = |r_\alpha - r_{\alpha'}| \tag{4.31}$$

and end up with a set of algebraic equations of the type

$$\Psi(\alpha,n) = \Psi_i(\alpha,n) + \sum_{\alpha'} \sum_{n'=0}^{n} G(\alpha,\alpha':n-n')\Psi(\alpha',n') \qquad n = 0, 1, 2, 3, \dots \quad (4.32)$$

where $\Psi(\alpha,n) = \Psi(r_\alpha, n\Delta t)$. This is our basic time-stepping algorithm. We can make clearer the formal solution method by rewriting this equation with terms involving the unknown $\Psi(\alpha,n)$ on the left-hand side as

$$\sum_{\alpha'} W(\alpha,\alpha')\Psi(\alpha',n) = \Psi_i(\alpha,n) + \sum_{\alpha'} \sum_{n'=0}^{n-1} G(\alpha,\alpha':n-n')\Psi(\alpha',n') \quad (4.33)$$

where we have introduced a weighting matrix W defined as

$$W(\alpha,\alpha') = \delta_{\alpha\alpha'} - G(\alpha,\alpha':0) \qquad (4.34)$$

This matrix contains two terms: the first is the delta function corresponding to elements of the unknown function appearing *outside* the integral sign; and the second are called "self-patch" terms and arise from the influence of the integral equation on the locality of the unknown space–time value.

One of the problems of the integral equation methods is the need to carefully evaluate these self–patch terms as they are usually singular, by which we mean that they contain an integral over terms such as $1/r$ when $r = 0$. Despite the fact that these singular points require care, we can integrate *around* such points, and so we say that the singularities are "integrable". The above considerations mean that we must use an analytical solution for the integral around singular points rather than applying the numerical approximations used elsewhere.

Given that we can determine these self-patch terms, we can then evaluate the weighting matrix and solve the above set by defining an inverse matrix W^{-1} such that

$$\sum_{\alpha'} W^{-1}(\alpha,\alpha')W(\alpha',\alpha'') = \delta_{\alpha\alpha''} \qquad (4.35)$$

and the formal solution is obtained as

$$\Psi(\alpha,n) = \sum_{\alpha'} W^{-1}(\alpha,\alpha')\left\{ \Psi_i(\alpha',n) + \sum_{\alpha''} \sum_{n'=0}^{n-1} G(\alpha,\alpha':n-n')\Psi(\alpha'',n') \right\} \qquad (4.36)$$

This equation forms the basis for the marching-in-time algorithm. The field at each instant n is obtained as the sum of the incident field at time n and the weighted sum of previously calculated values of the field. It is important to realize that evaluation of the inverse matrix W^{-1} is often trivial. This is due to the fact that usually $G(\alpha, \alpha' : 0) = 0$ if $\alpha \neq \alpha'$. In this case the weighting matrix reduces to diagonal form

$$W(\alpha,\alpha') = \left[1 - G(\alpha,\alpha{:}0)\right]\delta_{\alpha\alpha'} \qquad (4.37)$$

The inverse is simply found by inverting the scalar quantities $[1 - G(\alpha,\alpha{:}\ 0)]$.

A major problem with the time-stepping procedure is the accumulation of errors that can occur. We see that by replacing the integrals by discrete sums we are only approximating the integral equations and so (small) errors are introduced at the start of the procedure. However, since the evaluations involve weighted summations over all previous values, these small errors can grow and cause instability of the solution. The source of these errors has been carefully studied and there are now methods for the stabilization of such methods. It is important to realize however that long-time calculations can be difficult with time-marching methods due to the accumulation of these errors.

As an example of a time-domain integral method, Appendix B, Section 5 lists the MATLAB code for finding the current distribution on a wire antenna of circular cross-section. Details of the derivation of this equation can be found in the paper published by Liu & Mei (see Suggestions for further reading at the end of this chapter). Here we note only that the final integral equation is solved numerically using the time-stepping procedure outlined above. It is a version of the EFIE since it forces the electric field to be zero on the cylindical surface of the wire.

This code was used to find the current at the feed point of a wire antenna for comparison with the approximate methods used in Chapter 2. Comments are placed throughout the code to identify the various stages of the calculation.

The main output is the space/time current on a wire of length L. If we excite an antenna at the centre point with a Gaussian pulse with width much shorter than the transit time for light to propagate the distance L, we can view the current pulses move along the wire and be reflected from the ends in the qualitative manner described in Chapter 2.

Note that these are waves of *current* on the wire and are obtained by solving an integral equation when we know only that the feed point is being driven by a Gaussian pulse, that the total field on the surface of the wire must be zero and that the field can be expressed by a convolution integral involving a Green's function and current source term. The time-marching algorithm based on a retarded solution to the wave equation is then used to solve for the current on all segments of the antenna for all time.

An example of the feed point current obtained from this code for Gaussian voltage excitation is shown in Figure 4.11. Note that the current pulses decay as a function of time: this is due to radiation loss from the ends of the antenna and not to absorption in the material of the antenna (in deriving this model, we have assumed the wire material has infinite conductivity). Note also that the pulses spread as they propagate up and down the wire. This is due to dispersion, i.e. to loss of energy at high frequencies in the pulse. We ignored this phenomenon when developing our simple model of charge acceleration in Chapter 2.

We see then that the integral equation does contain all the physics of radiation from accelerating charges, as discussed in the context of the wave equation in Chapters 1 and 2. It is the role of such codes to provide the precision required for engineering calculations. However, we have shown that it is equally important to have a clear physical understanding of the principles involved in the wave mechanisms. This latter feature can be obtained using the simple procedures outlined in Chapter 2. The reader must beware of placing blind faith in the computer models; as we have shown for discretization of the advection equation, they can be very good at providing the wrong answer.

4.6 Case study: input impedance of dipole antennas

In order to illustrate the application of numerical integral methods we consider again the problem of determining the input impedance of a centre-fed cylindical dipole of length L and wire radius r (see Fig. 2.17). If we consider the case where $L = 1\,\text{m}$, $r = 1\,\text{mm}$, and drive the antenna with a Gaussian pulse of FWHH $= 1\,\text{ns}$ we obtain from the integral equation code (Appendix B, Section 5) the current shown in Figure 4.11.

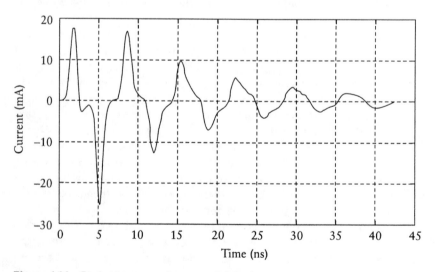

Figure 4.11 Feed point current for centre-fed dipole antenna.

Note that the source is assumed to have zero impedance and so the current pulses travel several times up and down the length of the antenna before decaying. If we measure the time between the first two pulses of current we see that the transit time is slightly longer than the 3.33 ns predicted if we assume the current waves travel with speed c. This is a real phenomenon, not an artefact of the numerical code, and is a consequence of the finite radius of the wire antenna (see Section 3.3).

Figure 4.12 shows the voltage drive point time history which we see is Gaussian in shape with a peak amplitude of 10 V.

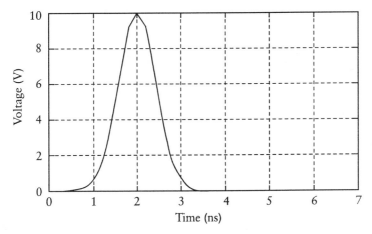

Figure 4.12 Feed point voltage excitation.

In order to find the input impedance we must first consider harmonic or sinusoidal excitations. By means of a Fourier transform (see Appendix C) we can obtain the spectrum of the drive pulse. This will show us the range of harmonic wave components present in the pulse (Fig. 4.13).

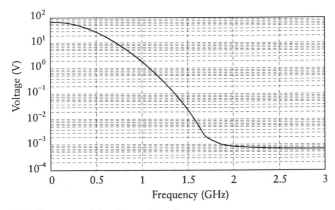

Figure 4.13 Spectrum of the drive point voltage.

This spectrum was obtained by using a fast Fourier transform (FFT) of the numerical drive point voltage. Similarly, we can find the current for a range of sinusoidal frequencies by taking the FFT of the current waveform. There results the spectrum shown in Figure 4.14.

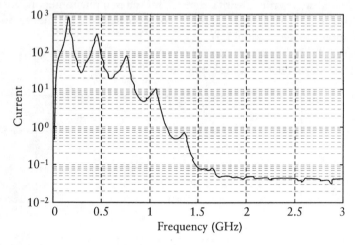

Figure 4.14 Spectrum of feed-point current.

We see that at very low frequencies the current falls to zero (as expected) but there are distinct resonances at frequencies for which the antenna length is a multiple of half a wavelength. These resonances are more clearly seen in the linear spectrum plot shown in Figure 4.15.

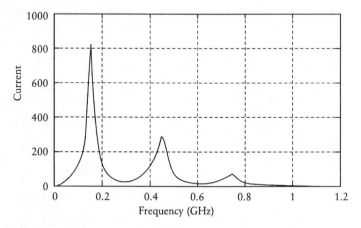

Figure 4.15 Linear spectrum of feed-point current.

We can then form the impedance as a function of frequency by dividing the voltage and current spectra at each frequency (Fig. 4.16).

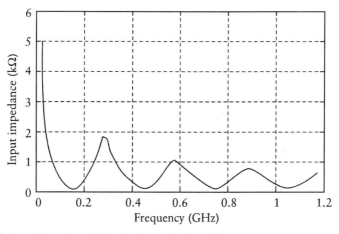

Figure 4.16 Input impedance of the dipole as a function of frequency.

Note the following points
- The input impedance is large at low frequencies, tending to infinity as $\omega \to 0$
- The impedance falls to a minimium around 150 MHz, the half-wave resonance (see Fig. 4.17). This minimum is around 73 Ω.
- The impedance shows a maximum of approximately 1.8 kΩ at a frequency of around 290 MHz.
- Both resonaces occur at frequencies slightly lower than predicted by the simple theory (150 and 300 MHz, respectively). This is due to the speed of propagation of the current pulse being slightly slower than c.

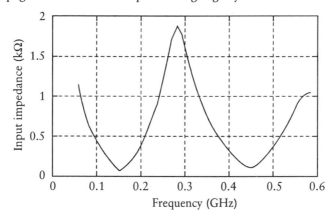

Figure 4.17 Input impedance at the half-wave and full-wave resonance.

Note that the impedance is a complex quantity (i.e. we can predict Z in amplitude and phase). Figure 4.18 shows the imaginary part of Z. Note how the impedance is large and negative at low frequencies, i.e. the antenna appears as a capacitive reactance at low frequencies. The physical origin of this is the predominant storage of energy in the electric field around the wire, the magnetic field and current being very small.

We can also use the variation of the imaginary part of Z to identify resonances. These occur when the imaginary part is zero. We can see from Figure 4.18 that this occurs at both the half- and the full-wave resonance, as expected.

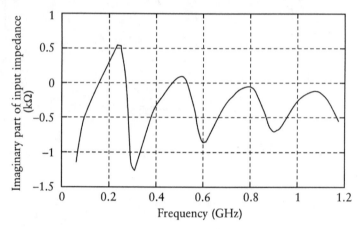

Figure 4.18 Imaginary part of the input impedance.

These values of impedance are more accurate and realistic than those based on the simple finite impulse response approximation outlined in Chapter 1 (which predicts values of zero and infinity for the impedance at half- and full-wave resonance, respectively). Nonetheless, we can see many of the general features of the simple model reflected in the functional form of the impedance variation. These general features are not so easily extracted from the original integral equation. In this way our simple models provide a means of checking the numerical codes and gives us insight into the important physical properties of the system.

In summary, we have seen that, with some care, numerical methods can provide the accuracy required for quantitative engineering design. However, we have also seen that discretization of Maxwell's equations is not straightforward and the numerical structure can introduce more problems than it solves. For this reason it is important to develop in parallel an understanding of the physical structure of the expected solution.

Suggestions for further reading

There are many new ideas introduced in this chapter and not many of them have yet been presented in published books. Therefore a list of original references have been given, where interested readers can find more details.

The following books on computational methods in electromagnetics are recommended.

Booton, R. C. 1992. *Computational methods for electromagnetics and microwaves.* New York: John Wiley.

Harrington R. F. 1968. *Field computation by moment methods.* New York: Macmillan .

Jones, D. S. 1987. *Methods in EM wave propagation,* vols I & II. Oxford: Oxford Science Publications.

Morgan, M. A. (ed.) 1990. *Finite element and finite difference methods in electromagnetic scattering,* (Progress in electromagnetics research, vol. 2. Amsterdam: Elsevier.

Ratnajeevan, S. & H. Hoole 1989. *Computer aided analysis and design of EM devices.* Amsterdam: Elsevier Press.

Silvester, P. P. & R. L. Ferrari 1990. *Finite elements for electrical engineers,* 2nd edn. Cambridge: Cambridge University Press.

Smith, G. D. 1985. *Numerical solution to partial differential equations: finite difference methods.* Oxford: Oxford University Press.

The following articles provide more details of topics discussed in the text:

Cangellaris, A. C., C. C. Lin, K. K. Mei 1987. Point matched time domain finite element methods for electromagnetic radiation and scattering. *IEEE Transactions* AP-35, 1160–74.

Fusco, M. 1990. FDTD algorithm in curvilinear coordinates. *IEEE Transactions* AP-38, 76–90.

Gomez Martin, R., A. Salinas, A. R. Bretones 1992. Time domain integral equation methods for transient analysis. *IEEE Antennas and Propagation Magazine* 34, 15–22.

Lynch, D. R. & K. D. Paulsen 1991. Origin of vector parasites in numerical Maxwell solutions. *IEEE Tranactions on Microwave Theory and Technology* MTT-39, 383–94.

Mieras, H. & C. L. Bennet 1981. Space time integral equation approach to dielectric targets. IEEE Transactions AP-30, 2–9.

Miller, E. K. 1988. A selective survey of computational electromagnetics. *IEEE Transactions on Antennas and Propagation* AP-36, 1281–1305.

Miller, E. K. & G. J. Burke 1992. Low frequency CEM for antenna analysis. *Proceedings of the IEEE* 80, 24–43.

Taflove, A. & K. R. Umashankar 1987. The FDTD method for EM scattering and interaction problems. *Journal of EM Waves and Applications* 1, 243–67.

There exists a method known as the transmission line matrix (TLM) method, which has many similarities to the time-domain finite difference method. It is fundamentally a discrete wave propagation system and has the same strengths and weaknesses as all such methods. Further details can be found in the following two articles, the first of which contains some supporting software for visualizing wave problems.

Hoeffer, W. J. R. & P. P. M. So 1991. *The electromagnetic wave simulator*. New York: John Wiley.

Johns, P. B. 1987. A Symmetrical Condensed Node for the TLM Method. *IEEE Transactions* **MTT-35**, 370–77.

Liu, T. K. & K. K. Mei 1973. A time domain integral equation solution for linear antennas and scatterers. *Radio Science* 8, 797–804.

Miller, E. K., A. J. Poggio, G. J. Burke 1973. An integro-differential equation technique for the time domain analysis of thin wire structures. 1. The numerical method. *Journal of Computational Physics* 12, 24–48.

Miller, E. K., A. J. Poggio, G. J. Burke 1973. An integro-differential equation technique for the time domain analysis of thin wire structures. 2. Numerical results. *Journal of Computational Physics* 12, 210–233.

Miller, E. K. & J. Landt 1980. Direct time domain techniques for transient radiation and scattering from wire antennas. *Proceedings of the IEEE* 68, 1396–423.

Mittra, R. 1976. Integral equation techniques for transient scattering. *Topics in Applied Physics* 10, 73–127.

Mohammadian, A. H., V. Shankar, W. F. Hall. Computation of EM scattering and radiation using a time domain finite-volume discretisation procedure. *Computer Physics Communications* 68, 175–96.

Moore, T. H. & J. G. Blaschek, G. A. Kriegsmann 1988. Theory and application of radiation boundary operators. *IEEE Transactions on Antennas and Propagation* **AP-36**, 1797–1813.

Mur, G. 1981. Absorbing boundary conditions for the finite difference approximation of the time domain electromagnetic field equations. *IEEE Transactions on EMC* 23, 377–82.

Rao, S. M., D. R. Wilton, A. W. Glisson 1982. Electromagnetic scattering by surfaces of arbitrary shape. *IEEE Transactions on Antennas and Propagation* **AP-30**, 409–418.

Rynne, B. P. & P. D. Smith 1990. Stability of time marching algorithms for the electric field integral equation. *Electromagnetic Waves and Applications* 4, 1181–205.

Rynne, B. P. & P. D. Smith 1991. Time domain scattering from arbitrary surfaces using the electric field integral equation. *Journal of Electromagnetic Waves and Applications* 5, 93–112.

Smith, P. D. 1990. Instabilities in time marching methods for scattering: cause and rectification *Electromagnetics*, 10, 439–451.

Yee, K. S. 1966. Numerical solution of initial boundary value problems involving Maxwell's equations in isotropic media. *IEEE Tranactions on Antennas and Propagation* **AP-14**, 302–307.

Problems

4.1 Use a finite difference aproach to generate a discrete version of the one dimensional wave equation (Eq. (4.17)). Prove that this method is stable as long as $\gamma < 1$ where $\gamma = v\Delta t/\Delta z$.

4.2 Use the leap-frog algorithm (Appendix B, Section 3) to investigate the pulse propagation when the relative permittivity of the medium is $\varepsilon_r = 1$, $\varepsilon_r = 4$ and $\varepsilon_r = 16$. Begin with a FWHH $= 10 dt$. Note how dispersion arises for high ε_r values. Investigate the effect of changing the FWHH to $5 dt$, $20 dt$ and $40 dt$. Explain your observations. Calculate the velocity of wave propagation from the space/time output and show that it agrees with the prediction of Maxwell's equations

4.3 Use the code given in Appendix B, Section 3 to investigate the effect of propagation in media with non-zero conductivity. Try setting $\sigma = 0.001$, 0.01 and 0.1. Examine the space/time plots and comment on the structure of the wave propagation.

4.4 Use the code given in Appendix B, Section 4 to estimate the reflection and transmission coefficients as a function of frequency through a slab with the following material constants:

    i $\varepsilon_r = 16$ $\sigma = 0.$
    ii) $\varepsilon_r = 16$ $\sigma = 0.001.$

Fix the slab thickness as 20 units and use the FFT statement to find the frequency dependent coefficients.

4.5 Use the code in Appendix B, Section 5 to investigate the effect of narrowing the pulse width in the estimation of current on a centre-fed dipole of length $L = 1$ m. Does the method become unstable for short pulse widths? Comment on the relative accuracy of the code predictions as a function of drive pulse width.

4.6 Modify the thin wire code given in Appendix B, Section 5 to feed the antenna from one end instead of from the centre. Using the ideas of Chapter 2, estimate the expected form of the current waveform and compare it with that obtained from the computer code, making careful comparisons of the timing of components in the impulse response.

4.7 Use Appendix B, Section 5 to calculate the input impedance of an end-fed wire antenna of length $L = 1$ m and radius 1 mm. Use the FFT in MATLAB to calculate the impedance as a function of frequency.

Chapter 5

Time domain radiation from aperture antennas

5.1 Aperture antennas

In Chapter 2 we showed how radiation occurs from accelerating charge. We saw how the principles behind such a physical phenomenon could be applied to the design of wire antennas such as dipoles and loops. In these cases we considered the conduction current flowing in metallic wires using a simple free-charge transport model.

In this chapter we consider a different but equally important class of antennas; those where the source is not directly due to accelerating charge, but to the establishment of electric and magnetic field distributions over a surface in space. These antennas are called "aperture antennas" since the field is generally zero over large portions of the surface and only non-zero over a limited region or aperture. We shall see that the behaviour of such antennas can be incorporated with our understanding of radiation physics by employing the concept of equivalent currents.

Examples of aperture antennas are the familiar parabolic dish antennas used for satellite communications and radar, as well as horn and slot antennas used for microwave communications.

The prototype aperture system considered here as an antenna is shown in Figure 5.1. Our basic problem is to find the field pattern for all z when the field distribution for $z = 0$ is known. Generally we shall assume that in the

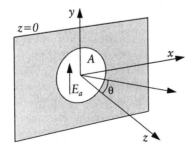

Figure 5.1 Aperture antenna geometry.

113

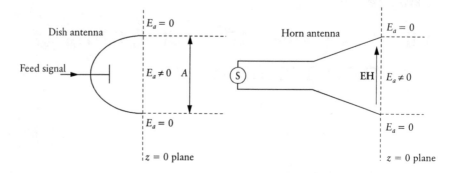

Figure 5.2 Examples of aperture antennas.

$z = 0$ plane the fields are zero except over some aperture or surface, shown as A in Figure 5.1. Such an assumption is a good approximation for the field pattern in the aperture of dish and horn antennas as shown schematically in Figure 5.2.

Note that we do not need to know how the field E_a is established; we care only that it has some prescribed value over the aperture A. This differs from the case we dealt with in Chapter 2 where we needed to know the conduction current in a wire before we could establish the radiated field.

The first point to note is that due to the phenomenon of *diffraction*, the field will not remain constant as a function of z. Energy will spread out from the aperture, the amount of spreading depends on the aperture size in terms of the wavelength of the field E_a. The smaller the aperture the larger the spread. We shall see how these concepts arise in the time domain for arbitrary excitations by considering the impulse response of the aperture considered as an antenna system.

The mathematical procedure we shall develop is based on Huygens' principle which, in its simplest form states that we can use knowledge of a wave across some surface to propagate the wave to all other points in space. Christiaan Huygens (1629–1695) was a Dutch physicist and astronomer who carried out pioneering work on wave propagation. This propagation is achieved by considering the known surface to be constructed from elementary point sources (Huygens' sources) which radiate into space according to the laws of wave propagation governed by the wave equation.

To formalize this procedure we have to consider an important new idea: the concept of equivalent currents. According to this approach we can replace our known field and aperture system by a set of *equivalent* current sources in Maxwell's equations (we shall see that one price we must pay for this equivalence is the need to introduce the idea of magnetic as well as electric currents in ME). Since we know how current sources radiate according to simple derivative rules, we can then establish the fields radiated from the aperture. The basis for these equivalent currents is Love's field equivalence theorem which we now consider.

5.2 Equivalent currents and boundary conditions

Consider the general field problem shown in Figure 5.3. Here we have two arbitrary volumes V_1 and V_2 bounded by surfaces S_1 and S_2. Inside V_1 are located all the sources of electric and magnetic fields, i.e. the electric currents $\mathbf{J}$ and electric charge of density ρ. These sources set up electric and magnetic fields $\mathbf{E}$ and $\mathbf{H}$ in space according to Maxwell's equations.

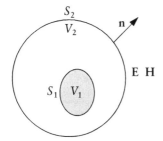

Figure 5.3 Surface S_2 enclosing the sources of an electromagnetic field inside V_1.

We consider a second surface S_2 which encloses V_1 and is specified mathematically by a surface normal vector $\mathbf{n}$ at each point on the surface. We want to establish fictitious currents over the surface S_2 such that the fields generated by these currents *outside* the surface S_2 are exactly the same as those produced by the original sources $\mathbf{J}$ and ρ. We consider the problem shown in Figure 5.4, where $\mathbf{J}_S$ are the fictitious currents which generate $\mathbf{E}$ and $\mathbf{H}$ outside the surface S_1 and generate different fields $\mathbf{E}_1$ and $\mathbf{H}_1$ inside the surface. We can determine the relationship between $\mathbf{E}$, $\mathbf{H}$, $\mathbf{E}_1$, $\mathbf{H}_1$ and $\mathbf{J}_S$ from the continuity conditions for fields tangential to a surface in space. These conditions

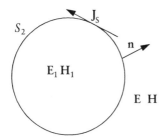

Figure 5.4 Boundary conditions across a surface S.

arise from the requirement that at all points in space the fields must satisfy the two Maxwell curl equations.

We consider application of the curl equation for the magnetic field H around a small rectangular contour which straddles a portion of a surface S. In the limit as the area of the rectangle tends to zero and using the definition of the curl as the limit of a contour integral (see Section 1.3), we arrive at a boundary condition which states that the surface current J_S must equal the magnitude of the discontinuity in tangential fields across the surface (Fig. 5.5).

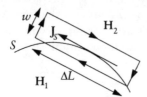

Figure 5.5   Boundary conditions across a surface S.

From the integral form of Ampère's Law we have

$$\oint_L HdL = J_S + \int_a \varepsilon \frac{\partial E}{\partial t} da \quad \Rightarrow \quad (H_{t1} - H_{t2})\Delta L = J_S \text{ as } w \to 0 \qquad (5.1)$$

where H_t are the tangential fields, and the contribution from the time derivative of the **E** field is zero as long as the derivative is bounded from infinity as we let w tend to zero (so that the area of the rectangular region tends to zero). We see that the Maxwell curl equations place constraints on the discontinuity of tangential field components across a surface. The key idea is that the magnitude of these discontinuities must be compensated for by surface currents J_S on the boundary. These are called "equivalent currents" since they are derived from field discontinuites, as opposed to "conduction currents" which arise from the movement of charged particles.

Since there are two Maxwell curl equations, we require two sets of equivalent currents: from Ampère's law we obtain the equivalent electric current source derived in Eq. (5.1). Applying this to Figure 5.4 we obtain

$$J_{es} = n \times (H - H_1) \qquad (5.2)$$

where J_{es} is termed an electric surface current since it derives from Ampère's law where the familiar conduction current also appears as a source of magnetic fields.

From Faraday's law, and following a similar argument to the above, we require a second kind of current source, which has no conduction equivalent, to model the discontinuity in the electric field. This is termed a "magnetic current" and is given by

$$\mathbf{J}_{ms} = -\mathbf{n} \times (\mathbf{E} - \mathbf{E}_1) \tag{5.3}$$

To be consistent with ME, a total equivalent surface current must then be defined as

$$\mathbf{J}_e = \mathbf{J}_{es} + \mathbf{J}_{ms} \tag{5.4}$$

The next stage of our procedure is to note that we can choose the fields $\mathbf{E}_1$ and $\mathbf{H}_1$ to be anything we desire. This may seem surprising, but follows from our initial requirement that these fictitous sources must only match the original fields external to the surface S_2, their effect inside S_2 is of no interest in the radiation problem.

It is convenient to select the internal fields to be zero, i.e. $\mathbf{E}_1 = \mathbf{H}_1 = 0$ everywhere inside V_2. In this case we see that our current sources are given simply by the values of the fields $\mathbf{E}$ and $\mathbf{H}$ over the surface S_2. But this is exactly the situation encountered in aperture antenna problems where we know the fields over some surface and want to find their values in some external space (outside V_2 in our example).

The final form of the equivalence principle states that we can replace any field problem by equivalent currents over a surface S, such that the equivalent currents and resulting fields are related as shown in Figure 5.6. This equivalence is totally consistent with Maxwell's equations in the sense that the fields *external* to S are unchanged by adopting these equivalent currents.

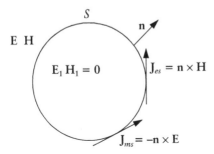

Figure 5.6 Equivalent electric and magnetic currents on a surface.

From our point of view we can now use these equivalent currents to solve for the radiated fields just as we did for the case where we considered only conduction currents, i.e. as an integral over retarded current derivatives. The only difference is that now we have to consider the Maxwell curl equations in a slightly more general form as

$$\nabla \times \mathbf{H} = \varepsilon \frac{\partial \mathbf{E}}{\partial t} + \mathbf{J}_{es}$$

$$\nabla \times \mathbf{E} = -\mu \frac{\partial \mathbf{H}}{\partial t} - \mathbf{J}_{ms} \tag{5.5}$$

where the extra term in Faraday's law (called a "magnetic current" since it is a source of electric field) stems from the equivalence principle.

Before proceeding to establish the radiated fields from such currents, we first show that we can modify this requirement to consider two types of current source so that in the calculation we need consider only one type. Since we chose the fields E_1 and H_1 to be zero it follows that we can replace the surface S by a material on the surface of which $n \times E = 0$. Such a material is called a "perfect magnetic conductor". In this case we short circuit the magnetic current J_{ms} which cannot radiate. Hence we have the situation that the fields E and H external to S can be found solely from the electric current J_{es} located on the surface of a perfect magnetic conductor. There is a price to be paid, however, since we must now consider the effect of this conducting surface on the remaining electric current.

Alternatively, we can use a material which is a perfect electric conductor (defined such that $n \times H = 0$ on its surface). In this case, we short circuit the electric current and the fields E and H can be found as the radiation from magnetic currents $J_{ms} = - n \times E$ placed on a perfect electric conducting surface. Using this technique we can always express the radiated field in terms of one kind of current only, provided we take into account the fact that the field must satisfy the boundary conditions for a perfect conductor of the appropriate type.

There are then three primary methods for calculating aperture radiation problems:

- Electric equivalent currents plus perfect magnetic conductor boundary conditions.
- Magnetic equivalent currents plus perfect electric conductor boundary conditions.
- Combined electric and magnetic current formulations radiating in free space.

To illustrate this, consider the calculation of currents on the aperture shown in Figure 5.1. If E_a and H_a are the known fields in the aperture A, then we generally assume that E_a is zero outside the aperture opening. The radiation field may then be found using one of the following three methods:

- Using the currents $J_{es} = a_z \times H_a$ and $J_{ms} = - a_z \times E_a$ where a_z is a unit vector in the z direction. These currents can be assumed to be radiating into a free space environment.
- Replacing the whole $z = 0$ plane with a perfect electric conductor and finding the fields radiated using *only* the magnetic current $J_{ms} = -a_z \times E_a$. In order to account for the effect of the perfect conductor on the magnetic current we must consider the influence of the current source plus its image (Fig. 5.7). We can account for the image by removing the surface and finding the total field from a magnetic current source of *twice* the magnitude, i.e. $2J_{ms} = -2a_z \times E_a$ radiating into free space.

- Similarly, using a perfect magnetic conducting surface and an equivalent electric current $2J_{es} = 2n \times H_a$ considered radiating into free space.

The possible configurations for imaging a current in a perfectly conducting plane are shown in Figure 5.7. Note that the polarity of the image arises from the requirement that the fields at the boundary satisfy the boundary condition for a perfect conductor.

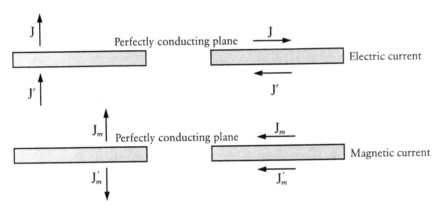

Figure 5.7   Images of currents in perfect conductors.

Note that throughout this discussion we have ignored the normal components of the field on the surface. This is justified from Maxwell's equations since if we ensure continuity of the tangential components it follows from Maxwell's equations that the normal components will be properly matched as well. For this reason we generally ignore the normal components when formulating aperture problems.

The success of these techniques relies on obtaining good estimates for the aperture fields. Usually we have to use approximations to obtain these fields and so the results for the three different current approaches can be different. It is important to realize, however, that this is not due to the basic equivalence theorem, but to the approximations used to find the currents. If we knew the currents exactly, all three methods would provide the same answer.

5.3 Wave guide propagation

As a simple example of the application of image theory to wave propagation consider the problem of determining the possible types or modes of propagation in a parallel plate metallic wave guide. The problem is shown schematically in Figure 5.8. A point source P generates a delta function between two parallel metallic plates with separation a. We suspect from simple optics that

119

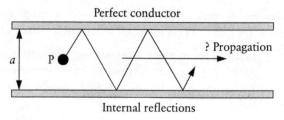

Figure 5.8 Wave guide propagation.

a wave can propagate by multiple internal reflections, as shown. The question is how do we model this using electromagnetic wave concepts?

We can use the image theory to replace this problem with an equivalent free-space problem. The only constraint is that we must generate fields which obey the boundary conditions of zero field along the plane sections of the plates. In this way we can remove the plates if we generate an infinite array of point sources of opposite polarity, as shown in Figure 5.9 (generated as images of P in the parallel "mirrors" formed by the metallic plates). P is the original source point with positive current. P1 is the image of P in the upper plate and has a negative current. Similarly, P2 is the image of P in the lower plate, P3 the image of P2 in the upper plate, and so on, to form an infinite array.

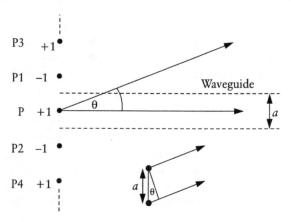

Figure 5.9 Array of point sources for waveguide propagation.

Note that the amplitude of the elements have alternating signs so that the fields cancel at the midpoints, as required by the boundary conditions. This system of radiating points is an example of an array antenna, i.e. an antenna constructed from a series of independent radiating elements with some relative position in space. Such antennas form an important class of radiating structures for many practical applications such as radio astronomy and radar imaging.

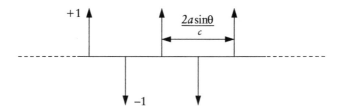

Figure 5.10 Impulse response of radiation inside a waveguide.

For the specific example considered it follows that the radiated impulse response at angle θ is given by an infinite set of alternating delta functions as shown in Figure 5.10.

From our discussion of antenna resonance (Chapter 3), we conclude that radiation from the array will only be efficient for certain discrete wavelengths (where the convolution shown in Fig. 5.10 matches the period of the exciting sinusoid). Since the impulse response contains an infinite number of elements, it follows that these resonances will be very narrow (see Figure 3.6). It follows that only certain discrete frequencies can propagate inside the guide (so-called "modes of propagation"). The lowest such frequency is called the "cut-off frequency" and is given when the period between the impulses in Figure 5.10 exactly matches one wavelength, i.e. when $\lambda = 2a\sin\theta$, which means that propagation occurs at an angle $\theta = \mathrm{asin}\,(\lambda/2a)$.

Since the maximum value of $\sin\theta$ is 1, it follows that there is a cut-off wavelength given by $\lambda = 2a$, above which no propagation is possible. (If we solve Maxwell's equations we can in fact show that an exponentially decaying wave is launched into the guide for frequencies below the cut-off frequency. This wave is called an "evanescent wave" and is rapidly attenuated to zero within a distance of a few wavelengths from the front of the guide.) This cut-off frequency is an important concept in the design of waveguides for communications and radar applications. For example, if we make a guide from two plates separated by 3 cm then no waves can propagate down the guide if they have a frequency less than 5 GHz. This effect can be used either to isolate the end of the guide from low frequency interference, or for propagating high frequency microwave signals from a transmitter to an antenna with low loss.

5.4 Radiation from equivalent currents

We showed in Section 5.2 that aperture radiation problems can be formulated using the concept of equivalent currents. We showed that such problems could be expressed in terms of either magnetic or electric currents, or both. We now consider how to find the electric field radiated by such currents.

For simplicity, we assume the presence of only one type of current denoted

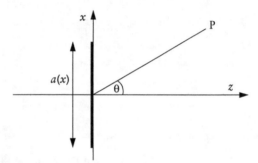

Figure 5.11 A linear aperture problem.

by $J(r,t)$ and start with our well-known relation between a current and radiated field in terms of an integral over retarded values of the derivative of the current:

$$E(\theta,t) = \frac{\mu_0}{4\pi r} \int_{-\infty}^{\infty} \frac{\partial J(x,\tau)}{\partial \tau} dx \qquad (5.6)$$

where $\tau = t - r/v$ is the retarded time variable and, for simplicity, we have assumed that the current is a function of only one space variable (x). The situation is shown in Figure 5.11, where $a(x)$ is some arbitrary function such that $a(x) = 0$ for $|x| > D$. If we now write the current as a product of two functions, one of space and the other of time, we obtain a general expression of the form:

$$J(x,\tau) = a(x)g(\tau) = a(x)g\left(t_0 + \frac{x\sin\theta}{v}\right) \qquad t_0 = t - \frac{R_0}{v} \qquad (5.7)$$

where $a(x)$ is our aperture distribution and $R_0 \gg D$ is the distance from the observation point to the centre of the aperture. We can now write the radiated field in the form:

$$E(\theta,\tau) = \frac{\mu}{4\pi R_0} \int_{-\infty}^{\infty} a(x) \frac{\partial g}{\partial \tau} dx = \frac{\mu v}{4\pi R_0 \sin\theta} \int_{-\infty}^{\infty} a(x) \frac{\partial g}{\partial x} dx \qquad (5.8)$$

If we integrate this expression by parts, we can transfer the derivative to the spatial distribution $a(x)$ to obtain

$$E(\theta,\tau) = \frac{\mu v}{4\pi R_0 \sin\theta} \int_{-\infty}^{\infty} a(x) \frac{\partial g}{\partial x} dx$$

$$\qquad (5.9)$$

$$= \frac{\mu v}{4\pi R_0 \sin\theta} \left\{ [ag]_{-\infty}^{\infty} - \int_{-\infty}^{\infty} \frac{\partial a}{\partial x} g \, dx \right\} = -\frac{\mu v}{4\pi R_0 \sin\theta} \int_{-\infty}^{\infty} \frac{\partial a(x)}{\partial x} g(\tau) dx$$

where the last stage follows from the fact that $a(x)$ is bounded in space and tends to zero as x tends to infinity. We shall see that this integral is easily evaluated when the derivative of the aperture distribution consists of delta function contributions, as it does for piecewise constant field approximations.

This is a very important result. It states that the radiated field from aperture antennas is given by the spatial derivative of the aperture distribution. As with the case of wire antennas, the sources of radiation are centred around discontinuities (such as edges) where the derivative is largest. To illustrate this point consider the example of radiation from a uniformly excited aperture such that (Fig. 5.12):

$$a(x) = \begin{cases} 1 & |x| \leq \dfrac{D}{2} \\ 0 & |x| > \dfrac{D}{2} \end{cases}$$

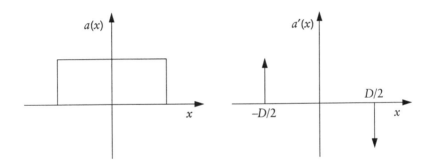

Figure 5.12 Uniform aperture distribution and its derivatives.

In this case the derivative is given by a pair of delta functions such that

$$\frac{\partial a(x)}{\partial x} = \delta\left(x + \frac{D}{2}\right) - \delta\left(x - \frac{D}{2}\right) \tag{5.10}$$

and the radiated field as a function of θ is found as

$$e(\theta,t) = -\frac{\mu v}{4\pi R \sin\theta}\left[g\left(t_0 + \frac{D}{2v}\sin\theta\right) - g\left(t_0 - \frac{D}{2v}\sin\theta\right)\right] \tag{5.11}$$

where $g(t)$ is the time variation of the current in the aperture. For delta function excitation we see that the radiated response depends on the angle θ and consists of two delta functions of opposite polarity and separated in time by an amount that depends on the angle θ.

In a direction normal to the aperture ($\theta = 0°$) there is no time separation between the two components and we obtain the special case of a radiated doublet D2 (see Fig. 3.2). This implies that the radiation *in this direction* is proportional to the derivative of the current excitation. In fact we can obtain a useful expression for the field assuming the impulse response is a doublet of width $\tau = D \sin(\theta/c)$. Using Eq. (3.3), we then have:

$$e(\theta,t) = \frac{\mu v}{4\pi R \sin\theta} \frac{D\sin\theta}{v} \frac{dg}{dt} = \frac{\mu D}{4\pi R} \frac{dg}{dt} \tag{5.12}$$

from which we see that the radiated field depends on the size of the aperture D as well as the time derivative of the excitation. Thus if we excite such an aperture with a step function we obtain in the far field a radiated impulse (Fig. 5.13).

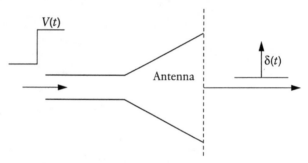

Figure 5.13   An aperture antenna as a differentiator.

This result is often used in time-domain electromagnetic measurement systems to obtain impulse waveforms for estimating impulse response. This is preferred experimentally, since it is easier to generate short rise-time step functions than other types of pulse function; in this scheme we use the antenna as a differentiator to obtain a radiated impulsive waveform.

A second important consequence of this result for aperture radiation is that, when establishing the field distribution across the aperture, we want as far as possible to achieve a uniform field distribution for efficient radiation. Very often the aperture will be the open end of a waveguide and if the waveguide is operating in some high-order transverse mode then there may be several spatial periods of the field across the aperture. In such a case, our acceleration rule tells us that we obtain partial cancellation in the far field and inefficient radiation. To stop this happening, mode converters can be employed to prevent high-order transverse modes developing. In advanced designs of aperture antenna systems, some of these higher order modes are deliberately developed to produce more desirable radiation patterns (for example, in aperture antennas for satellite communications where low levels of cross-polarization over a large bandwidth are required).

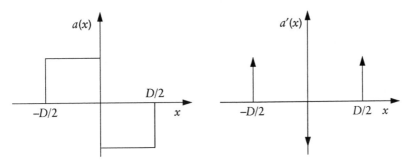

Figure 5.14 Piecewise constant aperture distribution and its derivative.

To illustrate the nature of this problem consider a piecewise constant aperture distribution which changes sign in the centre of the aperture, as shown in Figure 5.14. We see that on the axis we obtain approximately the second derivative of the time excitation $g(t)$. We could extend this idea to introduce more changes in sign across the aperture, and can see that this will lead to more and more delta functions of alternating sign with closer and closer time separations. For any given frequency the radiation would become smaller and smaller. Such over-moded patterns are then poor aperture distributions for efficient radiation.

5.5 Apertures as receiving antennas

As discussed in Chapter 2, the properties of an antenna are different when it is used as a transmitter and when it is used as a receiver. We showed that, in the case of a dipole antenna, the induced current at a point on the receiving antenna is given by the integral of the field radiated by the dipole in the driven transmitter case. This result also applies to aperture antennas.

In this case we have seen that an aperture with uniform illumination radiates the time derivative of the aperture distribution in the broadside direction. In the case of reception, we have a plane wave illuminating the aperture and want to know the resulting current (Fig. 5.15).

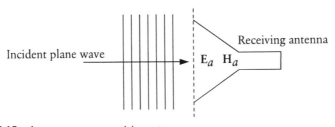

Figure 5.15 An aperture as a receiving antenna.

Using the integral rule outlined in Chapter 2, we expect the antenna to receive a faithful replica of the incident field, i.e. the received current impulse response is just a delta function D1 (or an integral of the radiated field doublet D2). This is indeed the case; horn antennas radiate approximately the time derivative of the drive signal, but receive a faithful replica of any incident fields. This important result can be used to obtain direct experimental estimates of the back-scattered impulse response of objects, as shown in Figure 5.16.

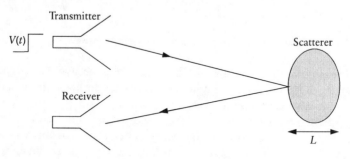

Figure 5.16 Principles of radar impulse response measurement.

The horn antenna is fed with a step function generator so the field radiated by the horn is an impulse (the time derivative of the step). For this reason the rise time of the step generator must equal the desired width of the radiated impulse waveform, which itself must be shorter than the time it takes for light to propagate across the principal dimension L of the scatterer. In this way we obtain a good estimate of the impulse response of the object (see Appendix A).

The incident field induces currents on the object. These currents re-radiate fields back towards the receiving horn. This process of induced currents re-radiating fields is termed *electromagnetic scattering* and any object which performs such a task is called a *scatterer*. The re-radiated fields generate a current in the horn which is a faithful replica of the back-scattered fields. In this way we can obtain a direct measurement of the impulse response of the scatterer. From this we can obtain the scattered response for arbitrary excitation by simple convolution.

The system we have just described is an elementary radar system ("radar" is an acronym standing for "radio detection and ranging"). Such radars can be used to locate objects in space and to track their movements. Although originally developed for military applications, radar has more recently been used for space-borne remote sensing, rain fall estimation and robotic sensing systems. In the next chapter we develop an important time-domain formulation of this back-scattering process which permits us to make direct estimates of the back-scattered field based on simple geometrical properties of the scattering object.

Suggestions for further reading

The following books and articles provide more details of the principles and key design elements of aperture antenna systems

Clarricoats, P. J. B. & G. Poulton 1977. High efficiency microwave reflector antennas: a review. *Proceedings of the IEEE* 65, 1470–1504.
Collin, R. E. 1985. *Antennas and radiowave propagation*. New York: McGraw Hill.
Ishimaru, A. 1991. *Electromagnetic wave propagation, radiation and scattering*. Englewood Cliffs, NJ: Prentice Hall.
Jones, D. S. 1987. *Methods in EM wave propagation*, volumes I & II. Oxford: Oxford Science Publications.
Kraus, J. D. 1988. *Antennas*. New York: McGraw Hill.
Lee, K. 1984. *Principles of antenna theory*. New York: John Wiley.

Problems

5.1 Find the equivalent magnetic and electric currents in the aperture of the open ended rectangular waveguide shown below, given that in the fundamental waveguide mode the fields across the aperture ($z = 0$ plane) are

$$E_y = E_0 \cos \frac{\pi x}{a}$$

$$H_x = YE_y$$

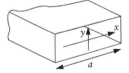

where Y is the mode admittance.

5.2 If the waveguide shown in Problem 5.1 is fed with a Gaussian pulse signal, i.e. $f(t) = \exp(-g^2 t^2)$, then write down the integral for the field radiated by the aperture, using only the equivalent magnetic current.

5.3 Calculate the shape and peak electric field strength radiated by a uniform aperture of dimension D in the broadside direction. The aperture is fed by a step function field with a rise time $t \ll D/c$, as shown below.

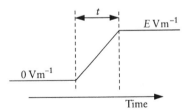

5.4 Repeat the calculations in Problem 5.3 but for radiation by the aperture in a direction $\theta = \pi/4$ from the broadside. Comment on the rise time of the radiated field compared to that of the broadside radiated pulse.

5.5 Write down a phasor expression for the field radiated by a uniform aperture as a function of θ when the aperture is excited by a sinusoidal signal $f(t) = \sin(\omega t)$. Find an expression for the peak field strength in the broadside direction. By normalizing the expression for arbitrary θ by the case for $\theta = 0$, show that the radiation pattern has the form of the sinc function defined as $\operatorname{sinc}(X) = (\sin X)/X$.

5.6 Show that the broadside radiation from the aperture distribution shown in Figure 5.11 is proportional to the second time derivative of the excitation field. Find the constant of proportionality and hence explain why such over-moded apertures are inefficient radiators.

5.7 A uniform aperture antenna of width D is used to receive a pulse signal at its broadside. If the incident pulse has the form shown below (where $t \gg D/c$), estimate the form of the received signal.

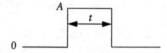

If the source is now moved so that the pulse arrives from an angle of 45°, estimate the form of the new received signal. Highlight the main differences between the two received signals.

Chapter 6
Electromagnetic wave scattering

6.1 The impulse response of a scattering body

In the last chapter we introduced the concept of equivalent currents and showed how we could explain radiation from aperture systems by considering an integral or sum over retarded derivatives of these equivalent currents. In this chapter we pursue these ideas a little further to prove the Kennaugh–Cosgriff formula relating the back-scattered impulse response of an object to the second derivative of its silhouette area function. This simple formula will allow us to predict the radar impulse response for a wide range of objects.

The basic problem we consider is as follows: a three dimensional object is illuminated by a plane electromagnetic wave which is launched from a transmitter at time $t = 0$ (Fig. 6.1). We assume the object is a perfect conductor, i.e. its conductivity is infinite (a good approximation for metallic objects at microwave frequencies).

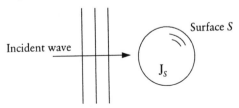

Figure 6.1 Scattering of plane waves by a surface S.

The incident wave generates electric currents on the surface of the body and these currents re-radiate back to the receive antenna, which measures the field and presents it to an observer. By timing the arrival of this return pulse, the range R of the object can be estimated directly, as shown in Figure 6.2. This is the basic principle of radar.

In the spirit of our discussion concerning wire antennas, we want to try and find a simple rule relating the form of the received signal to the geometry of the target, e.g. if it is a sphere of radius a or a square metal plate of side b. Such a rule does exist: it was first developed in 1965 by Kennaugh and Cosgriff, after whom the result is known.

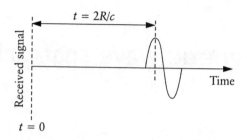

Figure 6.2 Radar location based on time measurement.

We begin by considering an integral relating the fields generated by changing currents on a body with surface S. We have

$$H_S(r,t) = \frac{1}{4\pi cR} \int_S \left\{ \frac{\partial J(r',\tau)}{\partial \tau} \times a_r \right\}_{\tau=t-R/c} dS' \qquad (6.1)$$

where H is the magnetic field strength at an observation point with position vector r and at time t. This equation contains all the physics of radiation considered earlier: radiation is caused by the current derivative at a retarded time τ (to establish causality). The cross product with the unit vector a_r is a mathematical statement of the transverse radiation law, i.e. only the components of the current derivative which are transverse to the direction of observation contribute to the radiated field. Finally, the factor $1/R$ models the decay of waves with distance, as discussed in Chapter 1.

Before we can find a relation between this radiated field and the object geometry, we need to find an expression for the currents $J(r',t)$ on a portion of the body. As mentioned in Section 4.5, finding these currents usually involves solving an integral equation (such as the electric field integral equation (EFIE) or the magnetic field integral equation (MFIE)). We can, however, make a simple but very important assumption for bodies whose local radius of curvature is always much greater than a wavelength. In other words, this is a high frequency theory, for objects which are electrically large.

The assumption amounts to approximating the *local* current on a body to be the same as that on an infinite plane tangential to the surface at that point. If we consider a scattering problem where the currents are induced by a plane wave incident on the body, then by invoking an electromagnetic boundary condition (see Section 5.2), this current must be equal to twice the tangential component of the incident magnetic field (assuming a perfect conductor), i.e.

$$J(r,t) = 2a_n \times H_i(r,t) \qquad (6.2)$$

where a_n is a unit vector normal to the surface at the point under consideration. Such an approximation is called the "physical optics" (PO) current.

Note that we are assuming that the object forms a perfect conductor so that we can ignore the magnetic current contribution. A similar formulation can be pursued for non-metallic objects if we introduce magnetic and electric current components. For simplicity we consider only the case where $\mathbf{J}$ is formed by an electric current.

The next stage is to choose a form for the incident magnetic field time variation in Eq. (6.2) and then to substitute in Eq. (6.1). Since we are interested in the impulse response we might begin by considering an impulsive or delta function variation. However, in anticipation of the final result, we will find it more convenient to consider an incident field which is a ramp function defined as

$$\mathbf{H}_i(\mathbf{r},t) = \mathbf{a}_h \, \mathrm{ramp}\left(t - \frac{\mathbf{a}_r \cdot \mathbf{r}}{c}\right) \qquad \mathrm{ramp}(\tau) = \begin{cases} \tau & \tau > 0 \\ 0 & \tau \le 0 \end{cases} \qquad (6.3)$$

where $\mathbf{a}_h$ is a unit vector in the direction of polarization of the incident field (Fig. 6.3). This ramp function is formally defined as the second integral of a delta function, a result which will be of prime importance in establishing the impulse response of our scattering body.

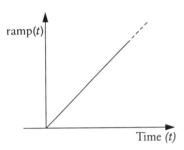

Figure 6.3   Ramp function.

If we substitute this function into the integral for the scattered magnetic field (Eq. (6.1)) we then need to evaluate two components of the resulting integral: a vector triple product, and the retarded time derivative of the incident magnetic field. We now consider each of these terms separately.

The vector triple product inside the integral has the form:

$$(\mathbf{a}_n \times \mathbf{a}_h) \times \mathbf{a}_r = (\mathbf{a}_r \cdot \mathbf{a}_n)\mathbf{a}_h - (\mathbf{a}_r \cdot \mathbf{a}_h)\mathbf{a}_n = (\mathbf{a}_r \cdot \mathbf{a}_n)\mathbf{a}_h \qquad (6.4)$$

where we have used the BAC–CAB rule to expand the triple product and the last step follows from the fact that $\mathbf{a}_r \cdot \mathbf{a}_h = 0$ since, as we saw in Chapter 2, electromagnetic waves are polarized transverse to the direction of propagation.

The time derivative is with respect to retarded time. The derivative of the ramp function is a step function $u(t)$ defined as (Fig. 6.4):

131

$$u(\tau) = \begin{cases} 1 & \tau > 0 \\ 0 & \tau \le 0 \end{cases}$$

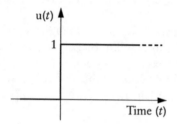

Figure 6.4 The step function as an integral of the ramp function.

When combined, these give rise to the following modified form for the radiation integral:

$$H_s(\mathbf{r},t) = \frac{a_b}{2\pi Rc} \int_S \left\{ u\left(\tau - \frac{\mathbf{a}_r \cdot \mathbf{r}'}{c}\right)_{\tau = t - \frac{\mathbf{a}_r \cdot \mathbf{r}'}{c}} \right\} \mathbf{a}_r \cdot \mathbf{a}_n \, dS' \tag{6.5}$$

We now note two important features of this integral: the first is that the scalar product $\mathbf{a}_r \cdot \mathbf{a}_n \, dS$ is dS_{proj}, the component of the surface area projected onto the line of sight, as shown in Figure 6.5.

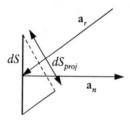

Figure 6.5 Projected area onto the line of sight.

The second important observation is that we can write the argument of the step function in terms of time t as:

$$\left(\tau - \frac{\mathbf{a}_r \cdot \mathbf{r}}{c}\right)_{\tau = t - \frac{\mathbf{a}_r \cdot \mathbf{r}}{c}} \equiv t - \frac{\mathbf{a}_r \cdot \mathbf{r}}{c} - \frac{\mathbf{a}_r \cdot \mathbf{r}}{c} = t - \frac{\mathbf{a}_r \cdot \mathbf{r}}{c/2} \tag{6.6}$$

Finally, we can write the radiation integral in the form

$$H_s(\mathbf{r},t) = \frac{a_b}{2\pi Rc} \int_S \left\{ u\left(t - \frac{\mathbf{a}_r \cdot \mathbf{r}}{c/2}\right) \right\} dS_{proj} \tag{6.7}$$

which simplifies to the following result, the Kennaugh-Cosgriff formula

$$H_s(\mathbf{r},t) = \frac{a_b}{2\pi Rc} A(t_s) \tag{6.8}$$

where $A(t_s)$ is the projected silhouette area of the scatterer. This function is defined as the time variation of the cross-sectional area of the scatterer in a plane transverse to the direction of propagation. This time variation is equivalent to that exposed by a wave travelling at a speed of *one-half* the speed of light.

This remarkable result states that the back-scattered ramp response field from an object is simply related to the silhouette area function of that object. Hence, if we know the shape of the object we can easily obtain an estimate of its back-scattered impulse response (and vice versa).

To find an approximation for the impulse response we need to differentiate this expression twice (since the ramp function is the second integral of the delta function) to obtain the response for a delta function illumination as the second derivative of the area function. Our final result for an approximation of the object's impulse response is then

$$\mathbf{H}_s(t) = \frac{a_b}{2\pi Rc} \frac{d^2 A(t_s)}{dt_s^2} \tag{6.9}$$

Note that this expression is for the back-scattered magnetic field, the electric field can easily be found from Maxwell's equations for plane waves as

$$\mathbf{E}_s(t) = Z_0 H(t) \mathbf{a}_v \tag{6.10}$$

where $\mathbf{E}$ is polarized at right angles to $\mathbf{H}$ and Z_0 is the impedance of free space defined as

$$Z_0 = \sqrt{\frac{\mu_0}{\varepsilon_0}} \approx 377\,\Omega \tag{6.11}$$

The full impact of this simple but useful formula is best appreciated by example. In the following we give two examples to illustrate the general method.

6.2 Example 1: Backscatter impulse response from a conducting strip

We consider the form of the back-scatter from a metal strip of width d illuminated by a plane wave incident at an angle θ as shown in Figure 6.6.

133

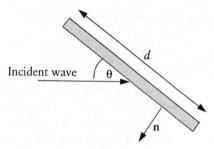

Figure 6.6 Scattering by a metal strip of width *d*.

In this case the silhouette area function is of the form shown in Figure 6.7.

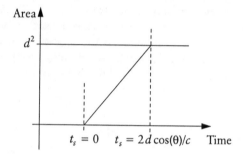

Figure 6.7 Area function for strip geometry.

If we differentiate this ramp function twice we obtain a direct estimate of the back-scattered impulse response from the strip as a pair of delta functions separated by the projected transit time between the edges (Fig. 6.8).

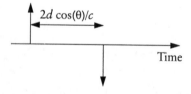

Figure 6.8 Second derivative of the strip area function: the physical optics impulse response.

We see that if the angle θ is $\pi/2$, that is normal incidence, we obtain the result that the back-scattered impulse response is given by a doublet D2, i.e. the back-scatter is proportional to the time derivative of the incident field (see Eq. (5.12)). In such a case we speak of a *specular* return. Such specular events are commonly experienced with the reflection of sunlight from polished metal or glass structures. Specular returns from flat metal plates are also used as radar calibration devices.

Note that even when $\theta \neq \pi/2$, when we might expect from Snell's law of geometrical optics that all the energy would be reflected away from the receiver, then according to physical optics we still obtain some back-scattered response.

For sinusoidal excitation this variation goes through maxima and minima as θ is varied (as we would expect from the fact that the impulse response contains two pulse components). The maxima of this variation are termed *sidelobes* and control of the level of such sidelobes is an important topic of design for reducing the back-scattered field from objects. Clearly, to modify such levels we must alter the impulse response which means we must change the current on the surface of the plate. We can do this either by changing the material properties or shape of the plate.

Figures 6.10–6.12 show the exact back-scattered field from a metal strip of width d illuminated at normal incidence, calculated using a finite difference time-domain solution to Maxwell's equation .

We consider two polarization configurations: transverse electric (TE) when the E vector is polarized transverse to the z direction, and transverse magnetic (TM) when the E vector lies along the z axis (Fig. 6.9).

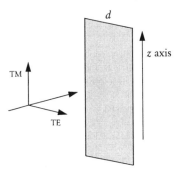

Figure 6.9 Geometry of finite difference calculations of back-scatter from a thin strip.

The metal strip is illuminated at normal incidence by a plane wave with Gaussian time variation. We show two examples: in the first the dimension d is small compared to the pulse width (Fig. 6.10) and in the second it is larger (Figs 6.11 & 6.12). Note the following important features:

- The TM case (Fig. 6.12) shows that back-scatter is proportional to the time derivative of the incident pulse as expected from our physical optics model
- The TE case is more complicated, where we see that the response has more time lobes.
- The location of the third time lobe in the TE response depends on the size of the strip.

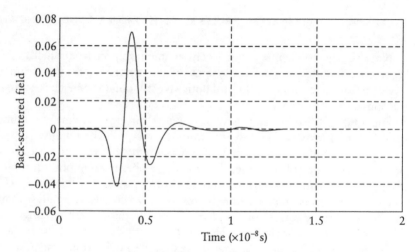

Figure 6.10 Back-scattered electric field for TE polarization of an electrically narrow strip.

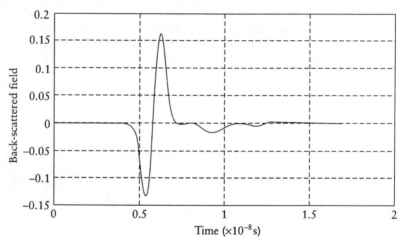

Figure 6.11 Back-scattered electric field for TE illumination of an electrically wide strip.

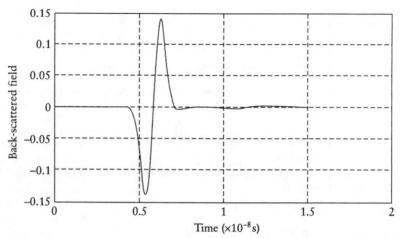

Figure 6.12 Back-scattered electric field for TM illumination of an electrically wide strip.

To resolve this discrepency for TE polarization we must note that the physical optics current is only an approximation to the true current. In actual fact, once currents are induced on a surface such as a flat metal strip, they are then free to flow, just as the current pulses on a wire antenna propagate as waves guided by the wire structure.

The point is that in deriving the physical optics model we have ignored these flowing currents on the surface and it is these that give rise to the extra time lobes shown in the exact solution above. But why do we see the effects of these currents for TE but not TM illumination? The answer is that for the former the currents are induced to flow across the strip and forced to zero at the strip edges, whereas in the latter case they are free to flow up and down the strip without a boundary, as shown in Figure 6.13.

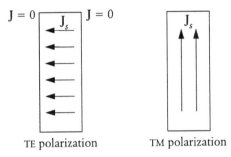

Figure 6.13 — TE polarization / TM polarization

Figure 6.13 Current induced on a strip by TE and TM polarization.

According to our understanding of radiation, we expect the TE currents to radiate (since charge is accelerating at the edges of the strip), whereas the TM currents will not (constant current so no radiation). This is indeed the case we observe in the radiated field. Clearly our physical optics current takes no account of these travelling currents on the body and so is valid only in the early time (or equivalently at high frequencies) before these travelling currents have time to propagate and radiate through charge acceleration.

We can obtain a better estimate of the back-scattered impulse response for the TE case by considering a more sophisticated model which generates current flow across the surface after the incident pulse has been applied. This current will not radiate until it hits the edge of the plate where charge deceleration will take place. If we assume that the charge moves across the plate with velocity c then this will cause a radiated pulse at a time d/c later. Our modified impulse response is then of the form shown in Figure 6.14, where we have defined the zero time origin as being at the centre of the plate. We could improve our estimate by adding further radiated pulses due to the current wave propagating between the edges of the plate. However, in practice, the radiation from the edges is so efficient that beyond the first such pulse the energy in the reflected current pulse is very small (unlike the case we observed for the dipole).

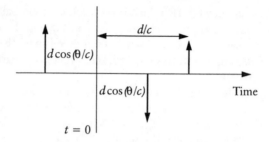

Figure 6.14 Modified impulse response approximation for a flat plate.

We can now see that if d is small and $\theta = \pi/2$, we obtain a triplet D3 for the impulse response, i.e. the back-scattered field is given approximately by the second derivative of the incident field (the delta functions have different relative amplitudes and so the triplet is asymmetric or lopsided). We observed such a triplet in the predicted radiated field for the TE case for a small plate (Fig. 6.10).

We see from this simple example how the physical optics approximation can provide good estimates for the early-time behaviour of back-scattered fields but that the late-time behaviour can require modifications to the surface current estimates.

6.3 Example 2: Backscatter from a metal sphere

The most symmetrical three-dimensional object is a sphere. Spherical scatterers are used as calibration objects in radar and, to a good approximation, microwave atmosperic scattering problems can be modelled as scattering from clouds of water particles that are spherical in shape.

The sphere is an important scattering problem because an exact solution to Maxwell's equations exists. This solution was first developed by Gustav Mie in 1910 and is expressed in terms of a mathematical series called the *Mie series solution*. It employs special functions of mathematics (called *Bessel functions*) to calculate the scattered field in any direction for a specific frequency. To find the time-domain response we must therefore evaluate this series at a range of frequncies and use an inverse Fourier transform (see Appendix C) to convert to the time domain. Figure 6.15 shows the variation in the magnitude of the back-scattered field from a metal sphere of radius a as a function of wave-number $k = 2\pi/\lambda$ when this exact Mie solution is used.

Note how the scattering variation with frequency has three important regions. For small k_a the scattering varies as the fourth power of frequency. Here the sphere radius is small compared to a wavelength and the behaviour is termed *Rayleigh scattering*, after Lord Rayleigh (1842–1919) who first

138

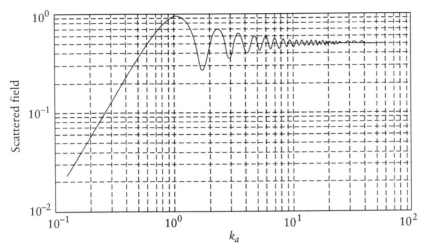

Figure 6.15 Mie solution for the backscatter from a metallic sphere, as a function of frequency.

studied this scattering phenomenon in an attempt to explain the scattering of sunlight by small particles in the Earth's atmosphere.

For values of k_a in the range $1 < k_a < 10$ we see that the response shows periodic fluctuations. This is termed the *Mie* or *resonance region*. As we shall see, this region is dominated by interference beween two components in the impulse response. In the limit $k_a \gg 10$, we see that the response becomes independent of frequency. This region corresponds to the optical region and to the case where the sphere radius is much larger than a wavelength. It is in this region that the physical optics current approximation is valid.

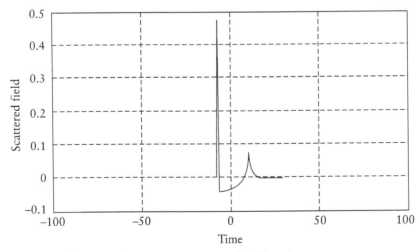

Figure 6.16 Sphere impulse response using a FFT of the Mie series.

139

Using this Mie series data we can use a Fourier transform to calculate the time-domain impulse response. This is shown in Figure 6.16 (where we have located the time origin at the centre of the sphere for convenience). The impulse response is limited in time and consists of three main features. The first is a strong specular flash from the front of the sphere. This is manifest as a delta function contribution to the impulse response. The second main feature is a smaller pulse contribution occurring later in time, which is due to what is called a "creeping wave" (Fig. 6.17).

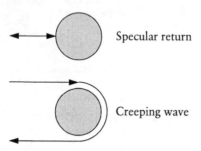

Figure 6.17 Scattering mechanisms from a sphere.

This is a current wave which flows over the surface of the sphere and re-radiates after transit around the circumference. Note that since the current is flowing in a curved path there will be continuous radiation from this current wave (in a similar manner to that observed for the loop antenna in Section 3.4). This radiation loss is the physical reason why it is reduced in amplitude over the initial specular flash. In theory, such waves continue to creep around the surface and provide periodic pulses in the impulse response. In practice, the radiation from such currents is so efficient that only the first such transit is seen in the impulse response.

These two pulse features are readily apparent from the calculated response shown in Figure 6.16, but there is a third important feature to the impulse response. We see that after the initial specular flash the radiated field does not return to zero but overshoots and becomes negative. This overshoot then decays into the creeping wave contribution. As we shall see, this component of the response is due to the physical optics current induced on the sphere.

We can obtain an approximation of the response for the sphere by using the Kennaugh–Cosgriff formula (Eq. (6.9)). The silhouette area function for a sphere is easily derived, as shown in Figure 6.18. The first and second derivatives of the area function are also shown. The impulse response is proportional to the second derivative and, according to physical optics, has the simple form shown in Fig. 6.18. Note that this response contains two of the features recognized in the "exact" Mie response (Fig. 6.16): namely, the initial specular flash from the front of the sphere, and the plateau of opposite polarity which is due to radiation by the physical optics currents as they are

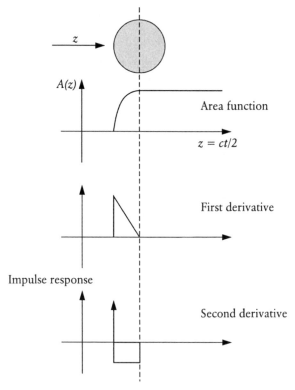

Area function

$z = ct/2$

First derivative

Second derivative

Figure 6.18 Physical optics impulse response for a sphere.

generated on the surface by the pulse moving over the sphere.

After a time $t = a/2c$ the physical optics response falls to zero. This occurs at the so-called "shadow boundary" which is defined as the limit of the maximum of the area function (in the case of the sphere it occurs at a distance of one radius from the front specular point). All points on the sphere beyond this are shadowed by the front surface, and so no currents are induced (Fig. 6.19). In reality, of course, we have seen that currents generated on the front surface can be guided as travelling waves into the shadow region and reappear as a creeping wave contribution. Physical optics takes no account of these currents.

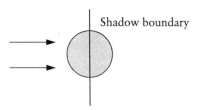

Shadow boundary

Figure 6.19 Shadow boundary for a sphere.

6.4 General constraints on the scattered impulse response

We conclude this section by making some general comments about the form of the impulse response for a general body. We will now show that whatever the form of the back-scattered impulse response $h(t)$, it must satisfy the following two global (integral) constraints:

$$\int_0^\infty h(t)dt = 0 \qquad \int_0^\infty th(t)dt = 0 \qquad (6.12)$$

Proof of these two important constraints follows from the physical fact that all radiated fields must be alternating, i.e. we cannot radiate a constant or d.c. field with an antenna of finite dimensions. If we could radiate such a constant field then from our radiation law it must be due to a current which is the anti-derivative of a constant i.e. an unbounded current, which is physically impossible.

The first constraint ensures that $h(t)$ has zero mean value, i.e. it has no d.c. component as required by our radiation law. The second constraint is less obvious, but relates to the fact that the step response must also have a zero d.c. component since it also must have a zero average. If it has a non-zero average then again we should expect an unbounded current response, which is not possible with a step excitation. Note that all higher moments (those like Eq. (6.12) but involving products like $t^n h(t)$) of the impulse response are not, in general, zero.

It is only these two moments which *must* be zero for all scattering bodies. (The next one in sequence is a ramp excitation. In this case we can have an unbounded current since a ramp, by definition, has unbounded amplitude and so the second moment need not be zero.) Our physical optics impulse response for the sphere satisfies the first, but not the second, of these two constraints (Fig. 6.18). With these constraints in mind we can determine the relative magnitude of the delta function and the plateau in the physical optics response, as shown in Figure 6.20.

The importance of these relationships is that they apply to the impulse response of any object and are not dependent on any particular shape or

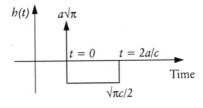

Figure 6.20 Quantified sphere impulse response using moment constraints.

form. They can be used to check the validity of a particular approximation or to devise simple modifications to enforce these constraints in approximate impulse response models.

Suggestions for further reading

The subject of electromagnetic scattering is usually treated as an advanced topic in electromagnetic theory. This means that many books available on the subject are difficult to understand without considerable mathematical sophistication. However, the following texts provide clearer than usual explanations of the physical processes involved, as well as covering in more detail the mathematical details of scattering calculations.

Hopcraft, K. I. & P. R. Smith, 1992. *An introduction to electromagnetic inverse scattering, Developments in Electromagnetic Theory and Applications*, 7, Dordrecht: Kluwer.

Ishimaru, A. 1991. *Electromagnetic wave propagation, radiation and scattering*, Englewood Cliffs, NJ: Prentice Hall.

Jones, D. S. 1989. *Acoustic and electromagnetic waves*, Oxford: Oxford Science Publications.

Kennaugh, E. M. & D. Moffat 1965. Transient impulse response approximations, *Proceedings of the IEEE*, 53, 893–901.

Ksienski, A. A., Y. T. Lin, L. J. White 1975. Low frequency approach to target identification, *Proceedings of the IEEE*, 63, 1651–1660.

Leonard Bennet, C. 1981. Time domain inverse scattering, *IEEE Transactions*. AP-29, 213–19.

Marx, E. 1987. Electromagnetic pulse scattered by a sphere, *IEEE Transactions*. AP-35, 412–17

Morgan, M. A. & B. W. McDaniel 1988. Transient electromagnetic scattering: data acquisition and signal processing , *IEEE Transactions on Instrumentation and Measurement*, 37, 263–7.

Tijhuis, A. G., 1987 *Electromagnetic inverse profiling: theory and numerical implementation* , Utrecht, The Netherlands: VNU Science Press.

Young, J. D. 1976. Radar imaging from ramp response signatures, *IEEE Transactions*. AP-24, 276–82.

Problems

6.1 Show that the back-scattered electric field from an object at range R is proportional to R^{-2}. Two identical scattering objects are placed at ranges of 30 m and 450 m from a radar. Estimate the dynamic range required in the receiver if we need to be able to detect the field from both objects. What is the maximum pulse repetition frequency of the pulse source if we wish to be able to measure unambiguous range out to $R = 500\,\mathrm{m}$?

6.2 Derive an expression for the radiated pulse width required to resolve two objects separated by a distance ΔR. If we want to design a radar to resolve objects which are only 10 cm apart, what maximum rise time of step generator can we use to feed a uniform aperture transmit antenna? If the peak drive step signal is 10 kV, estimate the field strength at a distance of 100 m from the radar.

6.3 Use the Kennaugh–Cosgriff formula to obtain an estimate for the back-scatter response of a finite cylinder of length L and radius a, as shown below.

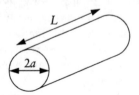

The axis of the cylinder lies along the x axis and the wave is normally incident on the cylinder (incident in the yz plane).

6.4 Sketch the form of the Radar impulse response of the body formed by rotating the following profile about the z axis by 2π.

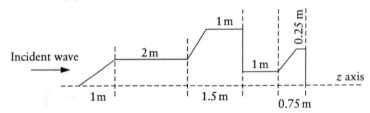

6.5 A metal plate lies in the xy plane with dimensions $2\,\mathrm{m} \times 1\,\mathrm{m}$ as shown below. This plate is illuminated at normal incidence by a Gaussian pulse of FWHH = 1 ns and polarized in the TE direction. Sketch the form of the scattered signal for back-scatter and for $\theta = 45°$. Compare the relative amplitudes of the two signals.

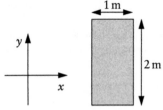

6.6 Repeat Problem 6.5 for the case of TM incident polarization. Comment on any differences observed.

6.7 Estimate the peak radiated field strength at 10 km from a uniform aperture antenna of dimension $D = 1\,\mathrm{m}$ when fed by a step pulse of 1 ns risetime and peak amplitude 10 kV. If this field is scattered by a metallic sphere of diameter 2 m, estimate the peak field strength observed back at the radar.

Appendix A

Introduction to time-domain measurements

We have seen that a time-domain description of electromagnetic waves and fields clearly exposes, through Maxwell's equations, the important physical properties of the origin of these waves. In this appendix we consider the principles behind the measurement of time-domain phenomena (see article by Baum in "Suggestions for further reading" at the end of this appendix), beginning with a description of time-domain measurements in sampled or discrete systems and then describing an important technique known as time-domain reflectometry (TDR).

A.1 Transient capture

The objective of sampling systems is to record a *faithful* replica of the analogue signal $f(t)$ as a sequence of digital samples. To illustrate, consider an ideal impulse sampling system, as shown in Figure A1, where we show how the process of sampling can be represented mathematically as the multiplica-

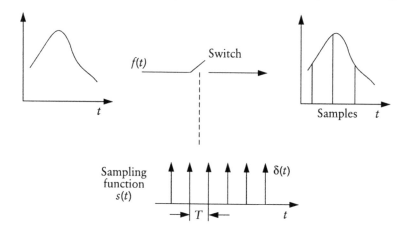

Figure A1 Idealized time sampling system.

tion of the function $f(t)$ by a sampling signal $s(t)$ which in this case is an infinite series of delta functions spaced by the sampling time T. We define a frequency $f_s = 1/T$ as the sampling frequency. Our basic problem is to determine how to set T in order that the samples contain all the waveform information present in $f(t)$. The answer is contained in a theorem known as the *sampling theorem*.

Clearly the time T must be sufficiently short to capture all the fine detail contained in $f(t)$ and must be much less than the overall duration of the signal. Just how much shorter depends on the nature of the short-time features in $f(t)$. Since such features are localised in the time domain it is difficult to establish a single general measure of $f(t)$ which we can use to set T. In this instance it is better to characterize the signal $f(t)$ in terms of its frequency spectrum. This spectrum $F(\omega)$ is defined from the Fourier transform (see Appendix C) of the signal $f(t)$ such that

$$F(\omega) = \int_{-\infty}^{\infty} f(t)\exp(-j\omega t)dt$$

$$f(t) = \frac{1}{2\pi} \int_{-\infty}^{\infty} F(\omega)\exp(-j\omega t)d\omega$$

(A1)

and we can describe the information in $f(t)$ in terms of its spectrum $F(\omega)$ as shown schematically in Figure A2.

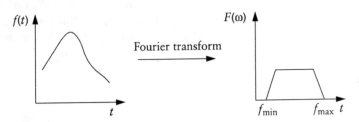

Figure A2   Spectrum of a signal.

The key feature is that the upper limit f_{max} will be dictated by the presence of short-time features in $f(t)$. In this way, instead of relating T to the duration of $f(t)$ we can relate the sampling frequency f_s to the maximum frequency in the signal spectrum f_{max}.

The second key idea we must now use is the convolution theorem (see Appendix C) which states that the Fourier transform of the product of two functions $f(t)$ and $s(t)$ is given by the convolution of their individual transforms $F(\omega)$ and $S(\omega)$. In order to use this result we note that the transform $S(\omega)$ is simply an infinite periodic array of delta functions separated by a frequency $\omega = 2\pi/T$, as shown in Figure A3.

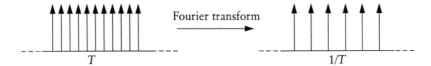

Fourier transform

Figure A3 Fourier transform of an idealized sampling waveform.

This is a very important result since we know that when we convolve a function with a delta function we simply obtain a replica of the function, hence it follows from the convolution theorem that the spectrum of a our sampled sequence $f(t)s(t)$ is a periodic array of spectra $F(\omega)$ centred on a grid with spacing $f_s = 2\pi/T$, as shown in Figure A4a.

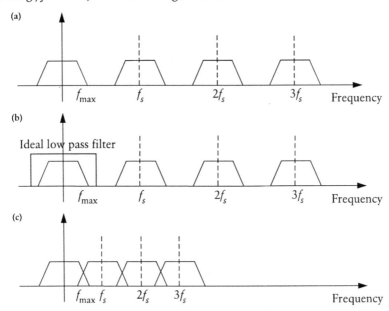

Figure A4 The sampling theorem and aliasing.

Figure A4a shows the case where $f_s \gg f_{max}$. In this case we see that the various spectral blocks are well separated. Remember that this is the spectrum of the sampled sequence and we wish to make sure that this sequence contains all the information present in the signal $f(t)$. We can achieve this by ensuring that there is no overlap in the spectral blocks. The condition for this is simply related to the ratio of f_s and f_{max}. As long as $f_s > 2f_{max}$ then there will be separation of the blocks and we can always completely reconstruct the signal $f(t)$ by passing the sampled sequence through a low-pass filter with a cut off frequency greater than f_{max}, as shown schematically in Figure A4b.

The criterion that $f_s > 2f_{max}$ is called the *Nyquist criterion* and is of fundamental importance in all sampled systems. Note that if it is not satisfied then the situation shown in Figure A4c arises. In this case there is overlap between the spectral blocks with the result that high frequency components of $f(t)$ can masquerade as low frequency components in the reconstruction of $f(t)$. Such distortion is called *aliasing*. A common example of such an effect is seen in films where spoked wheels often appear to move backwards. This is due to aliasing caused by the sampled film (generated by the camera shutter) and the fast rate of turn of the wheel. Aliasing can cause more serious effects in sampled electronic systems.

We now have a criterion for designing a transient digitizer for capturing the fast current and voltage wavefoms on antennas: we must provide a sampling rate of at least twice the highest frequency in our signal. In practice, this limit is a lower limit for the sampling of radio frequency (RF) signals and sampling rates of 5 to 10 times f_{max} are more often used. However, the Nyquist rate stands as a fundamental limit and demonstrates how difficult it is to design fast transient digitisers. If we wish to capture transient events on picosecond time-scales (in order to resolve fast transient currents on antennas) then we need a sampling system with a very wide bandwidth (of the order of several gigahertz). In practical terms, it is difficult and expensive to design such high-speed electronic circuitry which limits the use of such direct digitizers to relatively slow RF events (up to bandwidths of a few gigahertz). However, much wider bandwidths can be achieved using sampling scope technology, to which we now turn.

A.2 Sampling oscilloscopes

In our discussion of transient capture systems we saw that in order to sample a fast transient, we require very high-speed circuitry so as to satisfy the Nyquist criterion. If the signal $f(t)$ is a single transient event then we must face this problem directly, but if it is a repetitive waveform then we can achieve much higher effective bandwidths using the principle of sampling oscilloscopes.

Essentially such systems sample the waveform at successively later points in its cycle and then reassemble the samples into a replica of the signal. Since the input is repetitive, the sampling can be performed once every n cycles where n is some convenient large number. The point is that, although we still need to generate extremely narrow sample pulses, all amplification and signal conditioning can be designed at the much lower frequency of f_s/n. It turns out to be relatively easy to generate fast pulse generators but more difficult to obtain wide band amplifiers, and so such scopes can be made with very wide bandwidths.

The principles of such a sampling system are shown in Figure A5 and in

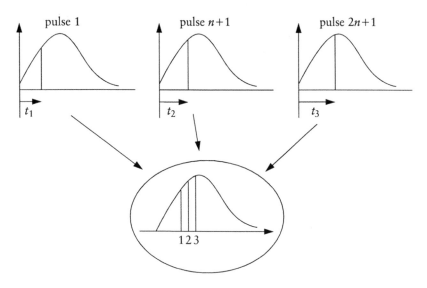

Figure A5 Principles of sampling oscilloscope operation.

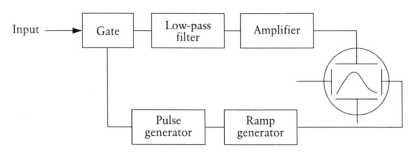

Figure A6 Key elements in wide band sampling system.

Figure A6 we show a simplified schematic block diagram of a sampling scope system. A wide range of such sampling scope systems are now commercially available, many with sampling times around 10–20 ps. These make possible direct capture of the fast current and voltage waveforms on antenna and scattering systems. In this way we can use the time domain approach to electromagnetics as the basis for an experimental investigation of the properties of electromagnetic wave structures.

A.3   Time-domain reflectometry

One important application of time-domain measurement techniques is time-domain reflectometry (TDR). In this technique pulses are launched as voltage

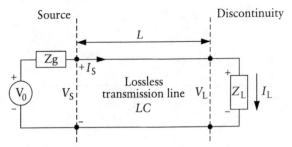

Figure A7 Voltage and current transmission line.

and current waves along transmission lines and any reflections from disconti-nuities are timed relative to the source point. The general situation is as shown in Figure A7. By knowing the velocity of propagation on the line we can calculate the location of the discontinuity causing the reflection (which may be a fault or terminating load impedances as shown in the figure). Fur-thermore, by examining the nature of the received signal, information can be gleaned about the nature of the discontinuity. In this section we outline the details of the TDR method as an example of the application of time-domain measurements.

The transmission line wave equation

The lossless transmission line is characterized by two parameters: the line inductance L (H m^{-1}) and the line capacitance C in (F m^{-1}). Note that these are not lumped circuit elements but are distributed parameters, so that over a short length dx the line has an equivalent circuit of the form shown in Figure A8.

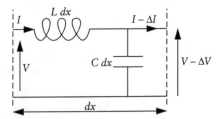

Figure A8 Circuit model for short section of line.

From Kirchhoff's voltage and current laws, it follows that:

$$(V - \Delta V) - V = Ldx\frac{\partial I}{\partial t}$$

$$(I - \Delta I) - I = Cdx\frac{\partial V}{\partial t}$$

(A2)

but, since dx is small, we can express the changes ΔV and ΔI in terms of partial derivatives as:

$$\Delta V \approx \frac{\partial V}{\partial x} dx$$
$$\Delta I \approx \frac{\partial I}{\partial x} dx \qquad \text{(A3)}$$

from which we obtain a small-scale or differential equation relating the space and time derivatives of current and voltage as:

$$\frac{\partial V}{\partial x} = -L \frac{\partial I}{\partial t}$$
$$\frac{\partial I}{\partial x} = -C \frac{\partial V}{\partial t} \qquad \text{(A4)}$$

which we recognize as a pair of coupled advection equations. It is then no surprise to learn that current and voltage waves can propagate on such a line. We can obtain a wave equation by differentiating the first equation with respect to x and the second with respect to t and substituting to obtain:

$$\frac{\partial^2 V}{\partial x^2} - LC \frac{\partial^2 V}{\partial t^2} = 0 \qquad \text{(A5)}$$

from which we see that the wave will propagate with a velocity

$$v = \frac{1}{\sqrt{LC}} \qquad \text{(A6)}$$

and that the general solutions for voltage and current are of the form of waves travelling in the $\pm x$ directions

$$V(x,t) = V_1\left(t - \sqrt{LC}x\right) + V_2\left(t + \sqrt{LC}x\right)$$
$$I(x,t) = I_1\left(t - \sqrt{LC}x\right) + I_2\left(t + \sqrt{LC}x\right) \qquad \text{(A7)}$$

If we now perform time differentiation of the current and space differentiation of the voltage, and equate using the advection equation, we obtain

$$\frac{\partial I}{\partial t} = \frac{1}{L}\left(-\sqrt{LC}\frac{\partial V_1}{\partial x} + \sqrt{LC}\frac{\partial V_2}{\partial x}\right) \qquad \text{(A8)}$$

and by integration we obtain a relationship between the current and voltage on the line as:

$$I(x,t) = \frac{1}{Z_c}\left[V_1\left(t - \sqrt{LC}x\right) - V_2\left(t + \sqrt{LC}x\right)\right] + A$$

$$Z_c = \sqrt{\frac{L}{C}}$$

(A9)

where Z_c is the characteristic impedance of the line. The constant A can be set to zero if we make the reasonable assumption that at time $t = 0$ all currents are zero (initial conditions).

If we now consider the situation at the load point then we have the following relationship between voltage and current at that point (boundary conditions):

$$R_L = \frac{V_L}{I_L} = \frac{V_1 + V_2}{(V_1 - V_2)/R_c}$$

(A10)

where the voltage wave components V_1 and V_2 are defined at the load point $x = L$. If we define a voltage reflection coefficient as $r = V_2/V_1$ then we obtain the following relationship between the load and the characteristic impedance:

$$\rho = \frac{R_L - R_c}{R_L + R_c}$$

(A11)

We see that the reflection coefficient is $+1$ if $R_L = \infty$ and -1 if $R_L = 0$. The basic design of a TDR system is shown in Figure A9. The pulse generator generates a step voltage with a rise time much less than the transmission line transit time T defined as $T = L/v$ where v is the velocity of wave propagation on the line. Usually the source impedance of the step generator is matched to the characteristic impedance of the line (which is often $50\,\Omega$).

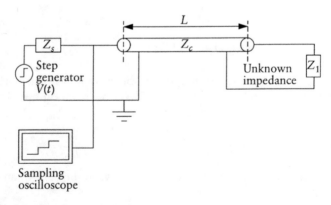

Figure A9 Principles of TDR measurements.

If the unknown load impedance equals Z_c then the pulse will be absorbed with zero reflection. However, if it is not equal to Z_c then some energy will be reflected back to the source, as described above. The reflected pulse will arrive at the TDR output after a time $2T$. The sampling scope captures the total waveform, given by the algebraic sum of the pulse generator pulse and any echoes.

Figure A10 shows a set of typical waveforms for resistive loads of various values.

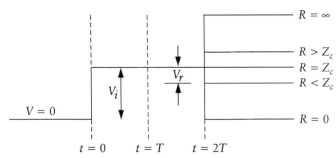

Figure A10   Typical waveforms from TDR measurements.

We see that the step response contains a plateau, the relative height of which depends on the load resistance (note that since we assume the source impedance equals the line impedance, there will be no more changes in the signal after time $t = 2T$). We can obtain a direct estimate of the reflection coefficient and hence load impedance by measuring the ratio of V_r to V_i as

$$\rho = \frac{V_r}{V_i}$$

$$R_L = Z_c \frac{1+\rho}{1-\rho} \tag{A12}$$

In the above analysis it was assumed that the load was resistive, but reactive loads can also be measured using this technique. Typical TDR response waveforms for inductive and capacitive loads are shown in Figure A11.

We see that the inductor initially appears as an open circuit to the fast rise time of the step. Similarly, the capacitor first reacts as a short circuit to the fast rise time. In the late time the inductor appears as a short circuit and the capacitor as an open circuit. By measuring the time constant t of the exponential behaviour between these two extremes, the values of L and C may be determined from:

$$L = \tau Z_c$$

$$C = \frac{\tau}{Z_c} \tag{A13}$$

153

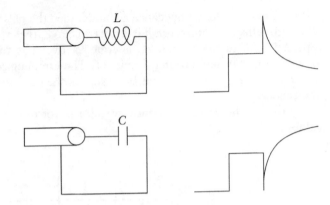

Figure A11   TDR response for reactive impedance loads.

These TDR techniques form a powerful set of experimental tests of impedance mismatch in high frequency systems, and are widely used in practice. They also provide the technology base required for the measurement of transient fields and currents established by wire and aperture antenna systems. In this way it is now possible to measure directly the impulse response properties of antenna systems.

Suggestions for further reading

Baum, C. E. 1976. Emerging technology for transient and broad band analysis and synthesis of antennas and scatterers, *Proceedings of the IEEE*, **64**, 1598–1616.

Appendix B
Computer codes for wave propagation

These 5 codes are all written using the matrix oriented programming language MATLAB as developed by MathWorks Inc. in the USA. This programming language is available for a wide range of computing platforms from PCs to Apple Macintosh and Sun workstations. Details of European price and availability can be obtained from the following address:

The MathWorks Europe Ltd.
Sheraton House
Castle Park
Cambridge
CB3 OAX
United Kingdom
Tel : 44 223 421 920
Fax : 44 223 421 921

The following codes make use of special matrix operations only available within MATLAB and so the reader is encouraged to run the programmes on a machine which has MATLAB available.

Note that these codes were written for MATLAB version 4.1, but do not require extra toolboxes for operation.

B.1 The one-way advection equation

```
function [z]=advect(n,nt);
% advect(n,nt) :
% advection equation for n spatial segments and nt time steps

u=zeros(1,n);% initial values of array set to zero
c=0.3;% speed of light in m/ns
dz=1/n;% space increment
dt=dz/c;% time increment
fwhh=10*dt;% width of Gaussian pulse
```

```
g=1.667/fwhh;
t=-dt;
si=2:n-1;% space index
clg;
fac1=c*dt/(2*dz);% scale factor for updating formula

for i=1:nt
    t=t+dt;
    ts=t-2*fwhh;% offset time
    v=exp(-g*g*ts*ts); % gaussian pulse excitation
    u(1)=v; %excitation placed on first element in the array
    u(si)=u(si)-fac1*(u(si+1)-u(si-1)); %updating formula
    plot(u);axis([0 n -3 3]);pause(0.1);
    z(i,:)=u;
end;
```

B.2   Lax modification to the advection equation

```
function [z]=advectlax(n,nt);
% advectlax(n,nt) :
% advection equation for n spatial segments
% and nt time steps using Lax averaging to stabilise

u=zeros(1,n);
c=0.3;%speed of light in m/ns
dz=1/n;
dt=dz/c;
fwhh=10*dt;%width of Gaussian pulse
g=1.667/fwhh;
t=-dt;
si=2:n-1;%space index
clg;
fac1=c*dt/(2*dz);% scale factor for updating formula

%Lax method
t=-dt;
for i=1:nt
    t=t+dt;
    ts=t-2*fwhh;
    v=exp(-g*g*ts*ts);
    u(1)=v;
    u(si)=0.5*(u(si+1)+u(si-1))-fac1*(u(si+1)-u(si-1));
```

```
    plot(u);axis([0 n -3 3]);pause(0.01);
    z(i,:)=u;% stores each set of space samples
end;

mesh(z,[45 80]); %plots matrix of space/time values
caxis([0 100]);
xlabel('space increment');
ylabel('time increment');
zlabel('field value');
```

B.3 The two-way wave equation

```
function [e] =wave2(ns,nt)
% wave2(ns,nt)
% one dimensional wave propagation
% ns is number of spatial segments (odd)
% nt is number of time steps
% bnd =0 is no boundary condition bnd=1 is first order boundary

dpt=round(ns/2);
bnd=input('request absorbing boundary 0=no 1=yes ');
d=1;%length of zone in m
t0=clock;%set clock to time length of computation
c=3e8;%speed of light
mu=pi*4e-7;%permeability of free space
eps=8.854e-12;%permittivity of free space
dz=d/ns;% space increment

er=input('input relative permittivity ');%choose relative permittivity of me-
dium
sig=input('input conductivity ');%choose conductivity
n1=sqrt(er);%refractive index
dt=dz/(c);%time step

fwhh=10*dt;%width of gaussian pulse
g=1.667/fwhh;%Gaussian pulse excitation
tb=2*fwhh;%offset time for Gaussian pulse

a=zeros(nt,2*ns+3);% set initial values of E and H to zero

%constants for use in updating formula
f1=dt/(dz*mu);
```

```
f2=sig.*dt ./(2*eps*er);
f3=(1-f2) ./(1+f2);
f4=dt ./(eps*er*dz.*(1+f2));
bnd3=0;

hindex=3:2:2*ns+1;%index for H field points
eindex=2:2:2*ns+2;%index for E field points

for t=2:nt
    ts=(t-1)*dt;
    %Gaussian pulse
    v=exp(-g*g*(ts-tb)*(ts-tb));
    drive(t)=v;

    %Hz calculation
    a(t,hindex)=a(t-1,hindex)+f1.*(a(t-1,hindex+1)-a(t-1,hindex-1));

    %Boundary Conditions
    if bnd==1
        a(t,1)=a(t-1,1)+(a(t-1,3)-a(t-1,1))/n1;
    else
        a(t,1)=a(t-1,1)+f1*a(t-1,2);
    end;
    if bnd==1
        a(t,2*ns+3)=a(t-1,2*ns+3)+(a(t-1,2*ns+1)-a(t-1,2*ns+3))/n1;
    else
        a(t,2*ns+3)=a(t-1,2*ns+3)-f1*a(t-1,2*ns+2);
    end;

    % Pulse Excitation of electric field
    a(t,2*dpt+3)=a(t,2*dpt+3)+v;

    %Ex calculation
    a(t,eindex)=f3*a(t-1,eindex)+f4*(a(t,eindex+1)-a(t,eindex-1));
end;

runtime=etime(clock,t0) %tells user the run time in seconds

x=1:2:2*ns+2;
h=a(:,x);%H field extraction
e=a(:,x+1);%E field extraction

%Display h field
mesh(h,[45 80]);% show H field
```

```
caxis([0 100]);
xlabel('space increment');
ylabel('time increment');
zlabel('field value');
```

B.4   Wave equation for a layer of thickness d

```
function [e,h] =layer
% layer
% one dimensional wave reflection from slab
% ns is number of spatial segments (odd)
% nt is number of time steps

ns=101;%default space segments
nt=300;%default time segments
perm=input('Input Relative Permittivity ');
cond=input('Input conductivity of medium ');

d=1;%length of zone in m
t0=clock;
c=3e8;
mu=pi*4e-7;
eps=8.854e-12;
dz=d/ns;
dt=dz/(c*1);
fwhh=10*dt;
g=1.667/fwhh;%Gaussian pulse excitation
tb=2*fwhh;

a=zeros(nt,2*ns+3);%set field values to zero
er=ones(1,2*ns+3);%set relative permittivity
sig=zeros(1,2*ns+3);%set conductivity

%set up slab thickness
half=round((2*ns+3)/2);
bnd1=half-20;
bnd2=half+20;
zone=bnd1:bnd2;

%set values for layer
er(zone)=er(zone).*perm;%relative permittivity of layer
sig(zone)=sig(zone)+cond;%conductivity of layer
```

```
n2=1;%free space on either side of layer
n1=n2;

f1=dt/(dz*mu);
f2=sig.*dt ./(2*eps.*er);
f3=(1-f2)./(1+f2);
f4=dt./(eps.*er.*dz.*(1+f2));
bnd3=0;
hindex=3:2:2*ns+1;
eindex=2:2:2*ns+2;

for t=2:nt
    ts=(t-1)*dt;
    v=2*exp(-g*g*(ts-tb)*(ts-tb));
    drive(t)=v;

    %Hz calculation
    a(t,hindex)=a(t-1,hindex)-f1.*(a(t-1,hindex+1)-a(t-1,hindex-1));

    %Boundaries + excitation
    a(t,1)=a(t-1,1)+(a(t-1,3)-a(t-1,1))/n1;
    a(t,2*ns+3)=a(t-1,2*ns+3)+(a(t-1,2*ns+1)-a(t-1,2*ns+3))/n1;

    %Ex calculation
    a(t,eindex)=f3(eindex).*a(t-1,eindex)-f4(eindex).*(a(t,eindex+1)-
        a(t,eindex-1));
    a(t,4)=a(t,4)+v;
end;

runtime=etime(clock,t0)%elapsed time in seconds

%extract E and H fields
x=1:2:2*ns+2;
h=a(:,x);
e=a(:,x+1);

%display h field
colormap('jet(64)');
a=abs(e);
a=a-min(a(:));
a=a/max(a(:));
image(a*64);
xlabel('space increment');
title('Plot of space/time H field');
```

ylabel('time increment');

B.5 Time domain integral equation for a wire antenna

```
function [stime]=hallen(L,r);
% hallen(L,r)
% Time marching solution to Hallen Integral Equation
% for current distribution on a stright wire of length L metres and
% circular cross section with radius r mm
%NOTE: this program can take a long time to run and requires a lot of
    memory
%As a starting point try L=1 r = 1 n = 32 nti = 128
%This should take about 2-3 minutes on a PC

c=3e8;%speed of light
zo=377;%impedance of free space
r=r*1e-3;%radius in m
ns=input('input number of space steps (even for centre feed) ');%number of
    space steps
nti=input('input number of time steps ');%number of time steps

Iz=zeros(ns+1,ns+1);%current space time array
forw=zeros(1,nti);%arrays for reflection from end of wire
back=zeros(1,nti);
alpha=zeros(1,ns+1);%arrays for integration of retarded current
beta=zeros(1,ns+1);

stime=zeros(nti,ns+1);%space time current array

si=L/ns; %space step
ti=si/c;% time step

ap=r/si;
lmin=L/r;
delta=si/(2*r);

%Assumes Gaussian pulse excitatin at feed point
time_step_nsecs=ti*1e9
disp('Note: For accurate results the gaussian must be at least 5 time steps
    wide');
fwhh=input('input width of Gaussian pulse excitation in nsecs ');
t0=clock;
```

161

```
fwhh=fwhh*1e-9;
g=1.665/fwhh;
amp=10;%amplitude of drive pulse
bias=10*ti;

%drive and observation points
dpoint=round((ns+2)/2);
obpoint=dpoint;

%Evaluation of current arrays
n=1:ns-1;
aa=sqrt((n+1).*(n+1)+ap*ap);
bb=sqrt(n.*n+ap*ap);
gamma1=(aa-bb)/(4*pi);
gamma2=log((aa+1+n) ./(bb+n))/(4*pi);

%Evaluation of self patch term

v1=pi*0.05635;v2=pi*0.443649;v3=0.25*pi;
ss=delta*delta;
fa=log(delta+sqrt(ss+sin(v1)^2));
fb=log(delta+sqrt(ss+sin(v2)^2));
fc=log(delta+sqrt(ss+sin(v3)^2));
c0=pi*(5*fa+8*fb+5*fc)/36;

fa=sqrt(ss+sin(v1)^2);
fb=sqrt(ss+sin(v3)^2);
fc=sqrt(ss+sin(v2)^2);
c1=pi*(5*fa+8*fb+5*fc)/36;

c0=(c0+(pi*log(2)/2))/(2*pi*pi);
c1=(c1-1)/(2*pi*pi);
c1=c1/delta;
c0=c0-c1;

za=4*zo*c0;

for q=1:nti     %time stepping begins
    for z=1:ns+1
        retard=bias-(q-abs(dpoint-z))*ti;
        s=amp*exp(-g*g*retard.*retard);
        v(z)=s/(2*zo);%calculates excitation for element z at time q

        %generates alpha array as integral over known currents
```

```
      extract=diag(Iz,z-1);
      f1=diff(extract)';
      sf1=length(f1);
      arr=f1(2:sf1).*gamma1(1:sf1-1);
      sindex=1:sf1-1;
      fac(sindex)=sindex.*f1(sindex+1);
      f2=extract(2:sf1)'-fac(1:sf1-1);
      arr=arr+f2.*gamma2(1:sf1-1);
      alpha(z)=sum(arr);

      %generates beta array as integral over known currents
      extract=diag(fliplr(Iz),ns+1-z);
      f1=diff(extract)';
      sf1=length(f1);
      arr=f1(2:sf1).*gamma1(1:sf1-1);
      sindex=1:sf1-1;
      fac(sindex)=sindex.*f1(sindex+1);
      f2=extract(2:sf1)'-fac(1:sf1-1);
      arr=arr+f2.*gamma2(1:sf1-1);
      beta(z)=sum(arr);
end;


v=v-alpha-beta;
% calculates reflection from end of wire
forw(q)=c1*Iz(2,2)-v(1);
if q-ns-1 >=0 forw(q)=forw(q)-back(q-ns); end;
Iz(1,1)=0;

%calculates reflection from end of wire
back(q)=c1*Iz(2,ns)-v(ns+1);
if q-ns-1 >=0 back(q)=back(q)-forw(q-ns); end;
Iz(1,ns+1)=0;

%current evaluation on each segment
for z=2:ns;
        if q-z < 0 f=0;
        else f=forw(q-z+1);end;
        if q+z-ns-1 < 0 b=0;
        else b=back(q+z-ns);end;
        Iz(1,z)=(v(z)-c1*(Iz(2,z-1)+Iz(2,z+1))+f+b)/(2*c0);
end;

%shift spacetime array and increment time step
p=ns+1:-1:2;
```

```
    Iz(p,:)=Iz(p-1,:);
    current(q)=Iz(1,obpoint);
    stime(q,:)=Iz(1,:);
end;

process_time=etime(clock,t0)

clg;
m=max(abs(stime(:)));
image(abs(stime)*64/m);colormap(jet(64));%view space time current
figure;
t=1:nti;
plot(t*ti*1e9,current);%plot current history at drive point
xlabel('time (nsecs)');ylabel('Current (amps)');
```

Appendix C
Transform techniques for solving EM wave problems

Throughout this book we have concentrated on a direct time-domain interpretation of electromagnetic wave phenomena. This has been a deliberate attempt to display as clearly as possible the physical properties of wave propagation and scattering. There is, however, an important alternative formulation of wave problems based on integral transforms which yields great analytical benefits and hence is widely used for quantitative engineering calculations. The techniques employed in these transform methods are well described in other textbooks and here we wish only to provide an introduction and to point out comparisons with time-domain methods

Transform techniques provide a reformulation of wave problems (cast in terms of the four-dimensional space–time (t, x, y, z)) in terms of a new set of variables; complex frequency and wavenumber (s, k_x, k_y, k_z) where $s = \sigma + i\omega \in \mathbf{C}$. The reason for employing such methods is that analysis in terms of these new variables is very often simpler than in the original space–time. However, an important constraint on useful transforms is the existence of the inverse operation, i.e. we must be able to "undo" the effects of the transform and return to the space–time domain. We then have a pictorial representation of general transform methods, as shown in Figure C1.

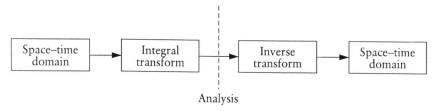

Figure C1   Generic form of integral transforms for wave problems.

A general integral transform has the following mathematical form

$$F(\alpha) = \int_a^b K(\alpha,t)f(t)dt \qquad (C1)$$

165

where the function K is called the *kernal* of the transform. The different possible transforms are generated by choosing different kernals and limits a and b.

While there are an infinite number of possibilities, the two most important such techniques are the Laplace and Fourier transforms which we now introduce in the context of wave motion.

C.1 The Laplace transform

In this book we have dealt with time signals of the form

$$g(t) = \begin{cases} 0 & t < 0 \\ f(t) & t \geq 0 \end{cases} \tag{C2}$$

A convenient way of representing such functions is in terms of the their Laplace Transform (LT) defined as

$$F(s) = \int_0^\infty f(t)\exp(-st)dt \qquad s = \sigma + i\omega \qquad \sigma,\omega \in R \tag{C3}$$

Note that there are certain conditions on $f(t)$ so that this integral, and therefore the transform exists. However, these conditions are less restrictive than those for the Fourier transform and so the LT exists for a wider range of functions.

The first thing to concern us is the existence of the inverse (see Fig. C1). Note that the transform maps a real variable (t) into the complex plane (s) and so the inverse must take us from the plane to the real line. This causes some complications and involves the calculus of complex variables. For our purposes, however, we note that the inverse may be formally defined (as a so-called *Bromwich integral*) but its realization is more complicated than the inverse Fourier transform (discussed later).

We represent the transform in Eq. (C3) as $F(s) = L(f(t))$ and the inverse as $f(t) = L^{-1}(F(s))$.

To illustrate the form of the LT, consider its application to some important time-domain signals. The transforms for the Dirac delta function, the step function, the exponential and the sinusoid are shown below. All these functions have been shown to be important in a time-domain description of antenna systems.

Example Laplace Transforms

Delta Function

$$f(t) = \delta(t - d) \;\Rightarrow\; F(s) = \int_0^\infty \exp(-st)\delta(t - d)dt = \exp(-sd)$$

Step Function

$$f(t) = u(t) \;\Rightarrow\; F(s) = \int_0^\infty \exp(-st)dt = \frac{1}{s}$$

Exponential Function

$$f(t) = \exp(kt) \;\Rightarrow\; F(s) = \int_0^\infty \exp(-st)\exp(kt)dt = \frac{1}{s - k}$$

Sinusoidal Function

$$f(t) = \sin(at) \;\Rightarrow\; F(s) = \int_0^\infty \exp(-st)\sin(at)dt = \frac{a^2}{s^2 + a^2}$$

The LT has two very important properties that make it useful for analysis:.

The LT of Derivatives of a function

If we define the LT of a function $f(t)$ as $L(f(t)) = F(s)$, then the following results hold for the LT of the time derivatives of f:

$$L(f') = sF(s) - f(0)$$
$$L(f'') = s^2 F(s) - sf(0) - f'(0)$$
$$\vdots \tag{C4}$$
$$L(f^n) = s^n F(s) - s^{n-1}f(0) - s^{n-2}f'(0) - \ldots - f^{n-1}(0)$$

where the initial conditions $f(0)$, etc., appear and are usually respresentive of
any energy sources which exist in the system at time $t = 0$. These results can
be proved by using integration by parts of the LT. The key observation is that
a differential equation involving time derivatives of a function will then be
replaced by a polynomial equation in s which is often simpler to solve.

We have seen that the wave equation and ME involve time derivatives, and so we can obtain new versions of these equations by applying the LT to the time variable, as shown below:

Laplace transform of Maxwell's curl equations

$$\nabla \times E(s) = -sB(s)$$

$$\nabla \times B(s) = \mu J(s) + \varepsilon \mu s E(s) = \left(\frac{\sigma}{s} + \varepsilon\right) s \mu E(s) \qquad (C5)$$

where we have used Ohms Law $J = \sigma E$, and σ is the conductivity of the medium. Note that we can now define the velocity of a wave using the coefficient of the time derivative of electric field (which will now be a function of σ) as

$$\upsilon = \frac{1}{\sqrt{\varepsilon'\mu}} = \frac{1}{\sqrt{\dfrac{\sigma\mu}{s} + \varepsilon\mu}} \qquad (C6)$$

We can thus define the complex velocity for a medium (complex because s is complex) with conductivity σ. Don't worry about trying to construct a physical interpretation of complex velocity. It turns out that such a concept combines the two phenomena of wave speed and attenuation through the complex exponential function, as illustrated below. Such strange ideas as complex velocity and phase shifts arise as a consequence of using the transform variable s. This ability to define mathematically the velocity in lossy media is a great strength of the LT and is one of its most important application areas in electromagnetics.

Using the derivative rules for the LT we obtain the following form of the wave equation

Laplace transform of the wave equation

$$\nabla^2 E(s) - \frac{s^2}{\upsilon^2} E(s) = 0 \qquad (C7)$$

where for simplicity we have assumed the initial conditions

$$E(x,0) = 0 \qquad \left.\frac{\partial E(x,t)}{\partial t}\right|_{t=0} = 0 \qquad \left.\frac{\partial^2 E(x,t)}{\partial t^2}\right|_{t=0} = 0 \qquad (C8)$$

To illustrate the consequences of this transformation, consider the case of the one-dimensional wave equation

$$\frac{d^2 E(s)}{dx^2} = \frac{s^2}{v^2} E(s) \tag{C9}$$

which is now an *ordinary* differential equation with solution in terms of the exponential function as

$$E(s) = c_1 \exp(-sx/v) + c_2 \exp(sx/v) \tag{C10}$$

(note that s is complex). The constants c_1 and c_2 are determined by the boundary conditions. Furthermore, if $E(s)$ is bounded at infinity then $c_2 = 0$.

The complex exponential function represents a decaying sinusoid (as shown in Figure C2) so we have made a connection between the Laplace variable s and the harmonic functions of Fourier analysis.

$$f(t) = \text{Re}(\exp(st)) = \text{Re}(\exp((\sigma + i\omega)t))$$

$$= \exp(\sigma t)\text{Re}(\exp(i\omega t)) = \exp(\sigma t)\cos(\omega t) \tag{C11}$$

Figure C2 shows an example plot of $f(t)$ for $s = -0.4 + i4$. From this solution it follows that if we know $E(t)$ at $x = 0$ we can obtain the solution at some other value of x by simple multiplication in the s domain and then inverse transformation, i.e.

$$E(0,t) = f(t) \implies L(f(t)) = F(s) \tag{C12}$$

$$L(E(x,t)) = \exp(-sv/x)F(s) \implies E(x,t) = L^{-1}(\exp(-sv/x)F(s))$$

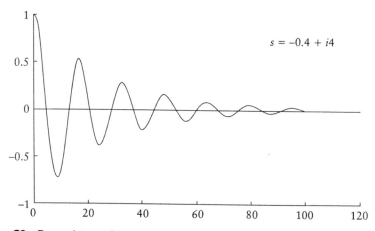

Figure C2 Damped sinusoid as represented by the complex frequency $s = \sigma + i\omega$.

In this way we are "propagating" the solution from one point to another by multiplying by the complex exponential function. This is to be contrasted with the need to solve the wave equation using finite differences in the time domain. There we need to propagate the field over the requisite number of time steps to obtain the solution for a new value of x.

This results in economy of effort in wave problems and is widely used to model wave propagation and scattering.

The convolution theorem (LT of products of functions)

A second very important property of the Laplace (and Fourier) transform is the convolution theorem which states that the transform of the convolution of two functions is given by the product of their individual transforms, i.e. if we define the convolution as

$$f * g = \int_0^t f(t - v)g(t)dv \tag{C13}$$

where the functions f and g have LTs $F(s)$ and $G(s)$ then we have

$$L(f * g) = F(s)G(s) \tag{C14}$$

We have seen that convolution integrals arise naturally in wave propagation and scattering, and so this theorem provides another strong incentive for using transform techniques in wave problems. It permits the evaluation of convolutions by simple multipication, and so is generally more efficient than the direct convolution evaluations considered in Chapter 2.

As a special case of this theorem we consider the case when the function f is the impulse response $h(t)$. In this case the LT of $h(t)$ is termed the *transfer function* $H(s)$ of the antenna or scatterer. It provide an alternative description of the system. We can then find the response of the system to any input waveform by multiplying the LT of the input by this function as shown in Figure C3.

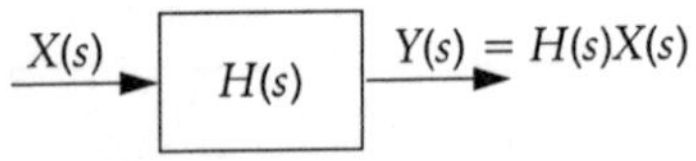

Figure C3 Transfer function based on the Laplace transform.

Note, however, that $H(s)$ must be known for all complex s and is itself a complex function having amplitude and phase. This makes it difficult to visualize the physical properties of the system from inspection of $H(s)$, especially when the impulse response contains delayed signal components due to wave propagation. These delays are manifest as phase shifts in the LT domain and so we must be careful to consider both the amplitude and phase of the transfer function when describing a system.

C.2 The Fourier transform

The Fourier transform (FT) is intimately related to the LT but has some very important differences that make it better suited to problems when we wish to transform with respect to spatial variables x, y and z. The transform may be derived as a limiting process from the well-known Fourier series or defined as a special case of an integral transform as

$$F(\omega) = \int_{-\infty}^{\infty} f(t)\exp(-i\omega t)dt \tag{C15}$$

where the parameter ω is real and to be interpreted as the angular frequency of a harmonic wave. In this way the FT is interpreted as a decomposition of the signal $f(t)$ into harmonic wave components with amplitude and phase given by the complex function $F(\omega)$.

Note that it appears that we can obtain the FT from the LT by replacing s by $i\omega$. While this is a useful concept, care must be exercised as the FT exists only if the above integral exists and the conditions on $f(t)$ for this to occur are more restrictive for the FT than they are for the LT (the extra $\exp(-\sigma t)$ appearing in the LT repesents a decaying exponential with increasing t and this can subdue any undesirable behaviour of the function $f(t)$ for large t, the FT has no such safeguard). As a result, some functions have a LT but do not have a FT.

Having said this, one benefit we obtain by using the FT is that the inverse is well behaved (it maps the *real* ω into a *real* t). It is defined as

$$f(t) = \frac{1}{2\pi} \int_{-\infty}^{\infty} F(\omega)\exp(i\omega t)dt \tag{C16}$$

As an example, consider the function $f(t) = u(t)\exp(-t)$, which is a decaying exponential step function. This function has a LT and a FT given by

$$L(f(t)) = \frac{1}{1+s} \qquad F(f(t)) = \frac{1}{1+i\omega} \tag{C17}$$

Consider, however, the function $f(t) = u(t)\exp(3t)$. This has a LT (for $\sigma > 3$), but the FT does not exist.

$$L(f(t)) = \frac{1}{s-3} \qquad F(f(t)) = \left[\frac{\exp(3-i\omega t)}{3-i\omega}\right]_0^\infty \qquad \text{(C18)}$$

The FT shares many of the useful properties of the LT, especially the convolution theorem and the rule for transforms of derivatives. However, it has two major advantages over the LT.

The first advantage is its ability to represent functions in the range $[-\infty, \infty]$. This means we can use the FT to transform functions of spatial variables x, y and z. The transform variable is then not a frequency ω but a wavenumber k.

$$F(k) = \int_{-\infty}^{\infty} f(x)\exp(-ikx)dx \qquad \text{(C19)}$$

Such transforms are widely used in wave problems since we can apply the FT in time *and* in space. If we transform Maxwell's equations using the FT, we obtain a set of algebraic equations of the form:

Fourier Transform of Maxwell's equations

$$\mathbf{k}.\mathbf{B} = 0$$

$$\mathbf{k}.\mathbf{E} = \frac{\rho}{\varepsilon}$$

$$\mathbf{k} \times \mathbf{B} = \left(\frac{\sigma}{i\omega} + \varepsilon\right) i\omega\mu\mathbf{E}$$

$$\mathbf{k} \times \mathbf{E} = -i\omega\mathbf{B}$$

where $\mathbf{k}$ is a wavenumber vector ($\mathbf{k} = (k_x, k_y, k_z)$) and we have again used Ohm's law to replace the current $\mathbf{J} = \sigma\mathbf{E}$. Note that all differential quantities have been removed from these equations. From elementary vector algebra it follows that in free space ($\rho = 0$), $\mathbf{E}$ and $\mathbf{B}$ are perpendicular to $\mathbf{k}$ and the magnitudes of $\mathbf{E}$ and $\mathbf{B}$ are related by a quantity called the "wave impedance", which is defined for a plane wave of frequency ω as

$$Z_\omega = \sqrt{\frac{\mu}{\varepsilon - i\frac{\sigma}{\omega}}} = a + ib \qquad \text{(C20)}$$

The second advantage of the FT is that the numerical evaluation of the forward and inverse FTs is easier than for the LT. In particular, when we consider sampled signals stored in computer memory, use is made of the discrete Fourier transform (DFT) in which we replace the integral of the FT by a finite summation. This DFT can be calculated numerically with great efficiency using a factorization of the summation in terms of complex exponentials with pure imaginary exponents. Such a factorization is not possible for the LT. This factorization is termed the fast Fourier transform (FFT). This algorithm has revolutionized the use of transform techniques in sampled data systems, including wave propagation systems. Many software packages (such as MATLAB as used in the codes given in Appendix B) have FFT routines already written, so users are not generally required to write their own. Such routines were used in Chapter 4 to calculate the impedance of the dipole.

Index